Development and Agroforestry

Scaling Up the Impacts of Research

Essays from *Development in Practice*

Edited by
Steven Franzel, Peter Cooper, Glenn L. Denning,
and Deborah Eade

A Development in Practice Reader

Series Editor
Deborah Eade

First published by Oxfam GB association with ICRAF in 2002.

ISBN 0 85598 464 3

A catalogue record for this publication is available from the British Library.

Available from:
Bournemouth English Book Centre, PO Box 1496, Parkstone, Dorset, BH12 3YD, UK
tel: +44 (0)1202 712933; fax: +44 (0)1202 712930; email: oxfam@bebc.co.uk

USA: Stylus Publishing LLC, PO Box 605, Herndon, VA 20172-0605, USA
tel: +1 (0)703 661 1581; fax: +1 (0)703 661 1547; email: styluspub@aol.com

For details of local agents and representatives in other countries, consult our website:
http://www.oxfam.org.uk/publications
or contact Oxfam Publishing, 274 Banbury Road, Oxford OX2 7DZ, UK
Tel: +44 (0)1865 311 311; fax: +44 (0)1865 312 600; email: publish@oxfam.org.uk

Our website contains a fully searchable database of all our titles, and facilities for secure on-line ordering.

The Editor and Management Committee of Development in Practice acknowledge the support given to the journal by affiliates of Oxfam International, and by its publisher, Carfax, Taylor & Francis. The views expressed in this volume are those of the individual contributors, and not necessarily those of the Editor or publisher.

Published by Oxfam GB, 274 Banbury Road, Oxford OX2 7DZ, UK.

Printed by Information Press, Eynsham.

Oxfam GB is a registered charity, no. 202 918, and is a member of Oxfam International.

Development and Agroforestry

Oxfam GB and ICRAF

Oxfam GB, founded in 1942, is a development, relief, and campaigning agency dedicated to finding lasting solutions to poverty and suffering around the world. Oxfam believes that every human being is entitled to a life of dignity and opportunity, and it works with others worldwide to make this become a reality.

From its base in Oxford, UK, Oxfam GB publishes and distributes a wide range of books and other resource materials for development and relief workers, researchers, campaigners, schools and colleges, and the general public, as part of its programme of advocacy, education, and communications.

Oxfam GB is a member of Oxfam International, a confederation of 12 agencies of diverse cultures and languages, which share a commitment to working for an end to injustice and poverty – both in long-term development work and at times of crisis.

For further information about Oxfam's publishing, and online ordering, visit www.oxfam.org.uk/publications

For further information about Oxfam's development and humanitarian relief work around the world, visit www.oxfam.org.uk

The International Centre for Research in Agroforestry (ICRAF), established in 1977, is an autonomous, non-profit research body supported by the Consultative Group on International Agricultural Research (CGIAR). ICRAF conducts strategic and applied research, in partnership with national research systems in the tropics, for more sustainable and productive land use. With its main offices in Nairobi, ICRAF collaborates with national centres and agencies in Indonesia, Mali, Peru, and Zimbabwe.

For further information about ICRAF, visit www.icraf.org

Contents

Contributors vii

Preface xi
Deborah Eade

Realising the potential of agroforestry: integrating research and development to achieve greater impact 1
Glenn L. Denning

Participatory design of agroforestry systems: developing farmer participatory research methods in Mexico 15
Jeremy Haggar, Alejandro Ayala, Blanca Díaz, and Carlos Uc Reyes

Participatory domestication of agroforestry trees: an example from the Peruvian Amazon 24
John C. Weber, Carmen Sotelo Montes, Héctor Vidaurre, Ian K. Dawson, and Anthony J. Simons

Facilitating the wider use of agroforestry for development in Southern Africa 35
Andreas Böhringer

Scaling up participatory agroforestry extension in Kenya: from pilot projects to extension policy 56
T.M. Anyonge, Christine Holding, K.K. Kareko, and J.W. Kimani

More effective natural resource management through democratically elected, decentralised government structures in Uganda 70
Thomas Raussen, Geoffrey Ebong, and Jimmy Musiime

On-farm testing and dissemination of agroforestry among slash-and-burn farmers in Nagaland, India 84
Merle D. Faminow, K.K. Klein, and Project Operations Unit

Scaling up the use of fodder shrubs in central Kenya 107
Charles Wambugu, Steven Franzel, Paul Tuwei, and George Karanja

The Landcare experience in the Philippines: technical and
institutional innovations for conservation farming 117
Agustin R. Mercado, Jr., Marcelino Patindol, and Dennis P. Garrity

Scaling up adoption and impact of agroforestry technologies:
experiences from western Kenya 136
Qureish Noordin, Amadou Niang, Bashir Jama, and Mary Nyasimi

Scaling up the benefits of agroforestry research: lessons learned
and research challenges 156
Steven Franzel, Peter Cooper, and Glenn L. Denning

Resources 171
    Books 171
    Journals 178
    Organisations 179
    Addresses of publishers 184

Index 186

Contributors

T.M. Anyonge is Natural Resources Programme Officer at the Swedish Embassy in Nairobi.

Alejandro Ayala works with INIFAP in Yucatán, Mexico.

Andreas Böhringer is a senior scientist with ICRAF, and has worked in projects related to natural resource management and rural development, largely in the Pacific Rim region and Africa.

Peter Cooper worked within the CGIAR systems for 21 years, both at ICARDA and then at ICRAF, where he led natural resource management (NRM) programmes. He now heads the Environment and Natural Resource Management Programme Area at IDRC, which supports research and development activities throughout Africa, the Middle East, Asia, Latin America, and the Caribbean.

Ian K. Dawson manages ICRAF's Genetic Resources of Agroforestry Trees Unit in Nairobi.

Glenn L. Denning is Director of ICRAF's Development Division, whose global, regional, and national activities focus on achieving impact and building capacity and institutional strength in agroforestry. His main professional interest is in bringing science and technology to bear on the development challenges of poverty, food security, and the environment.

Blanca Díaz previously worked with ICRAF and is based in Quintana Roo, Mexico.

Geoffrey Ebong is an agricultural economist currently working with the Association for Strengthening Agricultural Research in Eastern and Central Africa.

Merle D. Faminow was director of the Nagaland Environmental Protection and Economic Development Project (1998–2000), and is now a senior scientist with ICRAF.

Steven Franzel is a principal agricultural economist with ICRAF in Nairobi. Working in participatory technology development, he supports agroforestry research and dissemination teams throughout the tropics at sites where ICRAF works.

Dennis P. Garrity coordinates the research and dissemination activities of ICRAF in South-East Asia and leads its research in systems improvement in the region.

Jeremy Haggar conducts participatory research and training with small-scale coffee farmers through the Centre for Tropical Agricultural Research and Training (CATIE), and is based in Nicaragua.

Christine Holding is an ICRAF Research Fellow and was technical co-ordinator of the Nakuru and Nyandarua Intensified Forestry Extension Project. She is currently undertaking research in timber marketing from small-scale farms in Kenya.

Bashir Jama is a soil scientist who works on soil fertility and nutrient recycling, and is ICRAF's regional co-ordinator for East and Central Africa.

George Karanja is a forage agronomist working for the Kenya Agricultural Research Institute.

K.K. Kareko is Monitoring and Evaluation Officer at the Forestry Extension Services Branch of the Kenyan Forestry Department.

J.W. Kimani is Agroforestry Co-ordinator in the Soil and Water Conservation Branch of the Kenyan Ministry of Agriculture.

Kurt K. Klein is Professor of Economics at the University of Lethbridge.

Agustin R. Mercado, Jr. works on conservation-oriented agroforestry systems for sloping lands.

Jimmy Musiime is chair of Bubare Subcounty in Kabale District of Uganda. He is a teacher trainer and lecturer in environmental education at National Teachers College, Kabale.

Amadou Niang is ICRAF regional co-ordinator for the Sahel.

Qureish Noordin is a development facilitator at ICRAF's research station in Maseno, Kenya.

Mary Nyasimi is a field technician at ICRAF's research station in Maseno, Kenya.

Marcelino Patindol is president of the Claveria Landcare Association (CLCA) in Mindanao, in the Philippines.

Thomas Raussen works in Uganda with ICRAF and the Forestry Resources Research Institute.

Carlos Uc Reyes previously worked with ICRAF and is based in Quintana Roo, Mexico.

Anthony J. Simons heads ICRAF's Domestication of Agroforestry Trees Unit in Nairobi.

Carmen Sotelo Montes works for ICRAF on participatory domestication of agroforestry trees in Latin America, especially in the Peruvian Amazon Basin.

Paul Tuwei works at the Kenya Forestry Research Institute.

Héctor Vidaurre works for ICRAF on participatory domestication of agroforestry trees in Latin America, especially in the Peruvian Amazon Basin.

Charles Wambugu is dissemination facilitator at ICRAF, working on the CGIAR Systemwide Livestock Programme project on Utilisation of Forage Legume Biodiversity.

John C. Weber works for ICRAF on participatory domestication of agroforestry trees in Latin America, especially in the Peruvian Amazon Basin.

Preface

Deborah Eade

This *Development in Practice Reader* is based on an issue of *Development in Practice* commissioned and guest-edited by Steven Franzel, Peter Cooper, and Glenn Denning, all current or former[1] senior staff at the International Centre for Research in Agroforestry (ICRAF), which forms part of the Consultative Group for International Agricultural Research (CGIAR).[2] The volume comprises papers from their ICRAF colleagues around the globe, from the remote north-western region of India, to the Yucatán Peninsula on Mexico's Caribbean coast, from the hillsides of Uganda to the Peruvian Amazon.

Despite the somewhat more 'technical' or specialised focus than normally characterises titles in the *Readers* series, the authors not only convey their own passionate commitment to the small and often marginal farmers with whom they work, but also bring a depth of insight into wider debates that is fully grounded in their experience. Issues such as the relationship between theory and practice, the proper role of research in development, constraints on 'scaling up' (or, as one contributor calls it, 'scaling out'), local successes, the nature of human motivation for risk-taking and learning, and the ways in which individuals and communities respond to technical innovation, are all critically explored here. The value of learner-centred approaches is shown to be far greater than can be measured through the transfer of formal knowledge, and has as much to do with 'what works' as it does with any ideological principle. Participation and collaboration, for instance, may be good things in themselves or as a means to various ends, but the transaction costs of these approaches make it necessary not merely to invoke or romanticise such ideals, but also to demonstrate the tangible 'value-added' they bring to improving the situation of people living on the margins of the global economy. The conventional information- or technology-transfer model, based on 'simplifying the complex, separating the connected, and standardising the diverse', is shown to be misguided and wasteful. Contributors constantly stress the importance of exploring and experimenting with a range of possible agroforestry techniques and approaches to monitoring and evaluation, in conjunction with the farming communities who will adopt or reject these methods over time: however precarious their livelihoods, small and subsistence farmers are not interested in quick, but short-lived, fixes and indeed may well have a longer-term perspective than do people who can 'afford' to mortgage their futures. Again and again, the emphasis is on the importance of patience, and of tempering a commitment to social change with a willingness to be in it for 'the long haul'. Development agencies, which are accustomed to setting their own agendas and to re-fashioning them at will, would do well to heed what these highly experienced practitioners have to say.

Notes

1   Since the guest-edited issue was initially commissioned, Peter Cooper has left ICRAF and joined the International Development Research Centre (IDRC) at its head office in Ottawa.

2   ICRAF is based in Kenya. Other members of the CGIAR include CIAT (Colombia), CIMMYT (Mexico), CIP (Peru), ICARDA (Syria), ICRISAT (India), IFPRI (USA), IITA (Nigeria), ILCA (Ethiopia), and IRRI (Philippines). Further details are available in the resources section at the end of this book.

Realising the potential of **agroforestry:** integrating research and development to achieve greater impact

Glenn L. Denning

For more than two decades agroforestry has been heralded and actively promoted as a practical and beneficial land-use system for smallholders in developing countries. This promise led to the establishment of the International Centre for Research in Agroforestry (ICRAF) in 1978 and its support by the Consultative Group for International Agricultural Research (CGIAR) since 1991. Functioning initially as an information council during the 1980s, in 1991 ICRAF shifted its emphasis towards strategic research to strengthen the scientific basis for advocating agroforestry. This significant investment in process-oriented research greatly enhanced understanding of the opportunities and limitations of agroforestry and led to more critical assessments of its potential use (Sánchez 1995, 1999). As a result, agroforestry progressed from being an indigenous practice of great potential and romantic appeal to becoming a science-based system for managing natural resources (Sánchez 1995; Izac and Sánchez in press).

By the mid-1990s, the farm-level impact of agroforestry research was beginning to be observed in Africa and Asia. Much of this impact was a direct consequence of farmer-participatory research undertaken by ICRAF and its partners. Between 1992 and 1997, the number of farmers participating in on-farm research increased from 700 to more than 7000 (ICRAF 1998). Through such research, farmers acquired experience with the innovations, and this experience laid the foundation for pilot dissemination projects, and increased exposure to other farmers who did not directly participate in the research phase.

By 2000, several thousand smallholders in western Kenya were using short-term leguminous fallows and biomass transfer[1] to improve the fertility of depleted yet high-potential soils. In Embu District of eastern Kenya, more than 3000 farmers were planting tree legumes in fodder banks for use as an inexpensive protein supplement for their dairy cows. In Zambia, more than 10,000 farmers were using short-rotation improved fallows to restore soil fertility and raise maize crop yields. In the semi-arid Sahel region of West Africa, hundreds of farmers were adopting live hedges to protect dry-season market gardens from livestock. And in Southeast Asia, similar success was being observed on degraded sloping lands where hundreds of farmers in the southern Philippines were adopting contour hedgerow systems based on natural vegetative strips.

These examples from diverse ecoregions illustrate the emergence of sustainable agroforestry solutions to problems of land degradation, poverty, and food security in rural areas. The long-awaited promise of agroforestry as a science and as a practice is beginning to be realised at farm level. But impact on such limited scales, while certainly encouraging, cannot alone justify the millions of dollars invested in agroforestry research at ICRAF and national institutions over the past 25 years. Research institutions cannot rest on their laurels, having merely demonstrated that agroforestry has real potential. Instead, they must develop and implement strategies to ensure that millions of low-income farm families worldwide can capture the benefits of agroforestry.

This paper describes the approach that ICRAF has taken since 1997 to address the challenge of scaling up the adoption and impact of agroforestry innovations. To provide a conceptual foundation for scaling up, the first section provides a short overview of the literature and field experience regarding the constraints to adoption and impact. The next section describes institutional changes in the late 1990s that have embedded development within ICRAF's strategy, structure, and operations. These two sections form the basis of ICRAF's development strategy, which is outlined in eight focal areas of intervention and investment.

The fundamentals of adoption and impact

To increase the scale of adoption and the impact of innovations, action must be based on an understanding of the dynamics of adoption and the critical factors that determine whether farmers accept, do not accept, or partially accept, innovations. Adoption of agricultural innovations has been intensively researched since the seminal work of Grilliches (1957) on hybrid corn in the USA. Rogers and Shoemaker (1971) described adoption by individuals as an 'innovation-decision process', consisting of four stages as follows:

- *Knowledge*
  The individual is exposed to the existence of the innovation and gains some understanding of how it functions.

- *Persuasion*
  The individual forms a favourable or unfavourable attitude towards the innovation.

- *Decision*
  The individual engages in activities that lead to a choice to adopt or reject an innovation.

- *Confirmation*
  The individual seeks reinforcement for the innovation decision with the option of reversing that decision based on increased experience with the innovation.

The innovation-to-decision period is the length of time taken to go through this process, and it varies among individuals. Rogers and Shoemaker (1971) classified individuals by the length of their innovation-to-decision periods, categorising them as 'innovators', 'early adopters', 'early majority', 'late majority', and 'laggards'. This gave rise to the characteristic 'S' curve of cumulative adoption over time.

Schutjer and Van Der Veen (1977) noted that it is vital to consider the characteristics of alternative agricultural innovations when attempting to understand the importance of various constraints to adoption. One such characteristic is divisibility of technology. A divisible technology can be adopted to varying degrees. For example, innovations such as seed or fertiliser can be used across any proportion of a farm depending on the farmer's choice and resource limitations.

Low-income farmers are more likely to experiment with a divisible innovation because it can be initially tested on a small scale. Many agroforestry innovations are divisible and can be readily tested and evaluated by farmers in relatively small portions of the farm, such as along boundaries and in home gardens. Others, such as agroforestry for soil and water conservation, require an approach involving a whole farm, community, and watershed. This differentiation has important implications for scaling-up strategies.

Relatively few studies have explicitly examined the adoption of agroforestry innovations. Scherr and Hazell (1994) proposed a framework for analysing adoption from the perspective of a farming household. They divide the process into six sequential stages:

(1) knowledge of the resource problem, (2) economic importance of the resource, (3) willingness to invest long term, (4) capacity to mobilise resources, (5) economic incentive, (6) institutional support. Using this framework, Place and Dewees (1999) examined the effect of policy on the adoption of improved fallows, highlighting the importance of mineral fertiliser policy, production and distribution of planting material, and property rights. Franzel (1999) identified a number of factors that affect the adoptability of improved tree fallows. These were broadly grouped as factors affecting feasibility (such as the availability of labour, institutional support), profitability, and acceptability (perceptions of the soil fertility problem, past investments in soil fertility, wealth level, access to off-farm income). Franzel concluded that it is important to offer farmers different options to test, and to encourage them to experiment with and modify practices. The importance of farmer adaptation of innovations was also highlighted in a recent study on the adoption of alley farming in Cameroon (Adesina *et al.* 2000).

On the basis of relatively few empirical studies directly related to agroforestry, it is difficult to draw definitive conclusions about what are the most important factors affecting adoption and their implications for scaling up. However, drawing on the available literature, in particular the recent reviews of Franzel (1999) and Place and Dewees (1999), several factors are most likely to affect adoption of agroforestry innovations:

- biophysical adaptation of the innovation – the ability of the innovation to adapt and be adapted successfully to the farm environment;
- profitability of the innovation – in a broad sense to include consideration of returns to labour and land as well as financial profitability;
- farmers' awareness of the innovation;
- access to land, labour, and water;
- access to social capital, particularly where group action is needed;
- availability of essential inputs, particularly seed;
- access to financial capital;
- degree of risk and uncertainty.

Over the past decade, on-farm participatory research has played a crucial role in understanding and addressing the factors listed above. This approach has led to an increased role of farmers in diagnosing problems and in identifying and evaluating possible solutions. The result is better appreciation of farmers' perspectives and constraints, a more focused, farmer-centred research agenda, and, ultimately, higher levels of adoption (Franzel *et al.* in press).

The promotion and facilitation of innovation adoption amongst farmers are aimed at achieving positive impact. Yet the complexities of impact and the means to assess it are not well understood. The types of impact that result from adopting innovations can be broadly classified as economic, social, biophysical, and ecological, and are generally a combination of all four. To be more fully understood, impact has to be viewed from different spatial and temporal scales, as well as from the perspectives of different stakeholders (Izac and Sánchez in press).

Impact assessment is best undertaken through a framework that explicitly recognises the existence of trade-offs. For example, studies undertaken by the Alternatives to Slash-and-Burn Consortium in southern Cameroon demonstrated a clear trade-off between the global environmental benefits (carbon sequestration and biodiversity) and local profitability to farmers across a range of alternative land uses (Ericksen and Fernandes 1998). The research and development challenge is to understand the impact of adoption at these different scales (in this case, local versus global) and by different stakeholders (farmers versus the global community), and to optimise the trade-offs across a range of assumptions. Policy makers can then use this information to apply various policy instruments (for example, market intervention, land reform, infrastructure investments) that can affect the rate of adoption (Izac and Sánchez in press).

Impact over different temporal scales is an issue that is especially relevant to agroforestry in the developing world. Low-income farmers tend to discount heavily the potential long-term benefits of trees, opting instead for short-term practices that maximise food production and income. This slows the spread of soil conservation practices that have long-term benefits when the short-term effect on food production and income is negative (Fujisaka 1991). In contrast, farmers readily adopt agroforestry practices with short-term benefits such as short-term improved fallows (Kwesiga *et al.* 1999). The challenge for agroforestry research and development is to develop and introduce a range of options that provide an optimal trade-off between the long- and short-term expectations of farmers.

Institutional change: towards a research and development continuum

Now, after three decades of strong support to both international and national agricultural research, there are signs that growth has stagnated. Increasingly the call is for researchers to demonstrate the impact of past investments. This call is echoed at national levels where, in a climate of right-sizing in the public sector, ministries responsible for national budgets are starting to view public research as an extravagance. But the case for publicly funded research to address the challenges of food insecurity, poverty, and environmental degradation remains as compelling as it was in the 1960s. Research institutions must reinvent themselves to demonstrate that they are valuable and competitive investments of public resources. To this end, in the late 1990s, ICRAF embarked on institutional changes to foster and support greater impact of its research investments.

ICRAF's medium-term plan for 1998–2000 documented for the first time a clear institutional commitment to development impact (ICRAF 1997). The plan articulated three pillars of research: tree domestication, soil fertility replenishment, and policy, and two pillars of development: acceleration of impact, and capacity and institutional strengthening. In a departure from traditional CGIAR approaches to disseminating knowledge and technologies – that is, a reliance on networks, publications, and training as the principal vehicles of technology transfer – ICRAF and its partners adopted a more comprehensive and iterative functional model based on a continuum, from strategic research to applied research to adaptive research to adoption by farmers: a research and development continuum.

With this new approach, ICRAF and its partners accepted joint responsibility and accountability for ensuring the greater adoption and impact of agroforestry innovations. By proactively engaging in the development process, ICRAF could see four distinct benefits in institutional effectiveness:

- *Faster and greater impact* – by adopting a proactive rather than a passive approach to knowledge and technology dissemination, agroforestry innovations would reach more farmers, more quickly.
- *Innovation and learning* – by working directly and collaboratively with development partners in the field with farmers, opportunities would be greater for innovation and learning that would strengthen the knowledge and experience base of ICRAF and its scientists and thus share that asset with others.

- *A more relevant, demand-driven research agenda* – the innovation and learning associated with direct engagement in development would provide feedback to research on how innovations performed and generate hypotheses for future research.
- *Institutional credibility* – by demonstrating a clear commitment to greater impact of development, ICRAF would become a more credible partner in development and therefore could attract support from a broader group of stakeholders than would be the case if it assumed a strict 'research only' mandate.

In January 1998, ICRAF created a development division – the first of its type in the CGIAR system. The new division was established to complement the existing research division, which was responsible for planning and implementing an integrated natural resources management agenda related to agroforestry (ICRAF 2000; Izac and Sánchez in press). The development division brought together the existing development-oriented programmes and units of the centre: systems evaluation and dissemination, capacity building and institutional strengthening, and information. Both regionally and globally, the development division took on a more explicit, hands-on role in identifying, catalysing, and facilitating agroforestry-based opportunities for greater adoption and impact.

Integrating research and development activities at ICRAF takes place principally in the field in each of the centre's five operational regions: East and Central Africa, Southern Africa, the Sahel, South-East Asia, and Latin America. Strong regional leadership with an understanding and appreciation of the research and development continuum has been a major element of success thus far.

A second success factor has been the high level of 'buy-in' from the ICRAF board of trustees and from senior and middle management, including those individuals whose background and principal interest is research. After some initial concerns expressed about dilution of focus, lack of comparative advantage, and potential competition for limited resources, support and commitment have been strong. The understanding is clear that functioning through a research and development continuum actually strengthens support for research, and that greater field impact enhances the quality of scientific achievement. Both factors have been shown at ICRAF to have strong motivational effects on research scientists.

A third critical factor has been on-the-ground partnership with development organisations. ICRAF's comparative advantage has been, and remains in, applying science to development through agroforestry. Rather than trying to substitute for specialised institutions that have experience and expertise in development, ICRAF has sought to add value to their efforts through strategically focused interventions in development efforts in a partnership mode. From being a scientific leader in agroforestry with unique global knowledge and experience in integrating trees into farming systems and rural landscapes, ICRAF is now contributing importantly to the work of its development partners by providing technical support, training, and information, and by supplying seed.

An important issue to consider is whether the need for a development division within ICRAF will continue. The division has drawn interest and support during its first three years. In the longer term, however, it may be more appropriate that development becomes a mainstream way of doing business in much the same way that research on farming systems, environmental issues, and gender concerns have become mainstream in many research organisations after an initial period of special programme status.

Strategy for scaling up: crucial areas of investment and intervention

In September 1999, a two-day workshop at ICRAF brought together 23 national and international research and development specialists to discuss and identify the key elements of a successful scaling-up strategy (Cooper and Denning 2000). Drawing on seven case studies, participants identified 10 essential and generic elements (Figure 1).

Next, ICRAF sought to achieve greater adoption and impact by considering its institutional comparative advantage, using the outcome of this workshop, and referring to adoption literature. It devised a development strategy around eight areas of intervention and investment, as described below.

Policy makers

Public policy decisions can profoundly affect the uptake and impact of innovations (Place and Dewees 1999). The 1998 CGIAR system review (Shah and Strong 1999) highlighted the importance of policy research and dialogue in bringing about a better enabling environment. ICRAF is increasing efforts to facilitate and catalyse policy change through collaborative research and through formal dialogue with important policy and decision makers.

Source: Cooper and Denning 2000

Higher education institutions

ICRAF's success in both research and development critically depends on the capacity of partners: individuals and institutions. In 1993, ICRAF established the African Network for Agroforestry Education (ANAFE) as a collaborative mechanism for universities and colleges teaching agroforestry and related subjects. By 2000, ANAFE had 103 member institutions in 35 countries, becoming the largest network of education institutions in Africa (ANAFE 2000). The goal of ANAFE is to promote the institutionalisation of agroforestry in higher education institutions in order to produce graduates better equipped to develop, disseminate, and implement sustainable agroforestry and natural resource management practices. In 1998, a similar network was established in South-East Asia with 35 collaborating institutions.

Basic education institutions

Basic education institutions have enormous potential to expand the reach of agroforestry information and technologies. Building on related investment and experience by other institutions in environmental and health education, ICRAF has initiated a Farmers of the Future programme that aims to reach the next generation of farmers and, through them, to influence the current generation. The main areas of intervention will be education policy change, improvement of curricula and teaching resources, awareness creation, pilot projects linking schools and communities, and education systems research.

Seed supply systems

The lack of seed, seedlings, and other planting material is frequently identified as the most important constraint to greater adoption of agroforestry (Simons 1996). This shortage often disappoints farmers who must depend on relatively ineffective public and private sectors. ICRAF's focus in this area is to develop and apply better methods of forecasting germplasm needs, and to help establish effective, low-cost, sustainable, community-based systems of producing and distributing germplasm.

Community organisations

It is increasingly recognised that empowering local communities to control their own decisions and resources is fundamental to any successful development strategy (Binswanger 2000). A trend is emerging in developing countries towards devolving power to local government and increasingly to local communities. This devolution is coupled with building capacity in the community. ICRAF's experience with introducing and adapting the Landcare movement in the Philippines demonstrates the key role of community organisations in helping to scale up the adoption and impact of agroforestry innovations (Mercado et al. 2001). ICRAF sees a continuing role in catalysing and documenting institutional innovation through action-research with development partners. There is also a continuing need to develop and share relevant agroforestry innovations as entry points for community action.

Product marketing systems

Better markets for agroforestry products provide a way for poor farming households to generate income (Dewees and Scherr 1996). The key challenge is to improve the structure, conduct, and performance of agroforestry tree product markets and to make those markets accessible to low-income producers. ICRAF plays a role in placing market research in the mainstream of agroforestry research and development programmes, in developing innovative marketing methods, and in building marketing capacity. Another key role is that of a knowledge broker on aspects of agroforestry marketing, including aspects of processing that add value to products.

Extension and development organisations

Mainstream extension organisations and development institutions are often in a position to expand the reach of innovations. Extension contacts are particularly important during the early stages of farmer experimentation with innovations (Adesina *et al.* 2000). ICRAF is working closely with government extension systems, NGOs, and development projects to catalyse greater adoption and impact. The major contributions are in providing research support and technical advice, studying approaches to dissemination, and helping organisations to share their experiences.

Research institutions

Demand-driven, impact-oriented research institutions are needed to ensure a flow of innovations to rural areas. Yet frequently we find that research agendas are unresponsive to field realities and poorly linked to extension. Through collaboration, training, workshops, and publications, ICRAF has actively promoted participatory on-farm research approaches and the research and development continuum as a potential operational model for national research institutions, for the reasons elaborated earlier in this paper.

Through these focal intervention areas, ICRAF aims to reach 80 million agricultural poor by 2010, providing them with access to agroforestry options that improve livelihoods and sustain the environment (ICRAF 2000). ICRAF's development strategy is founded on strong partnerships and strategic alliances with a diverse group of institutions that share the Centre's mission and complement its expertise and reach.

Conclusions

In 1997, ICRAF set forth on a new and less-travelled path for an international agricultural research centre. Recognising that agroforestry research had the potential to deliver new livelihood options for reducing poverty, improving food security, and sustaining environmental quality, the centre unilaterally expanded its mandate to include a more proactive, hands-on approach to achieve greater impact.

ICRAF took this unconventional step because the impact of natural resources management research (including agroforestry) had in the past been limited and sporadic, suggesting that the traditional Green Revolution approaches to research and development were not universally appropriate. As we move beyond the food bowls of Asia to meet the challenges of more complex, heterogeneous, and often marginal environments, more site-, farmer-, and community-specific solutions are required. In order to improve our understanding of these circumstances, researchers need to be closer to policy makers and the more direct clients – smallholder farmers and the change agents who work with rural communities – to test, adapt, and share innovations. Because of this approach, ICRAF's research agenda has evolved in a way that is more relevant to the real needs of, and opportunities for, the agricultural poor.

By directly engaging in the development process through strategic partnerships with development institutions, the impact of research on food security, poverty reduction, and environmental sustainability will be realised more quickly and on a greater scale than with classic technology transfer approaches that use publications as the principal vehicle for disseminating research findings. Research institutions must therefore broaden their thinking and their mandates to the point where they can function as, and be seen as, credible development partners.

The developing world has no shortage of successful and well-publicised pilot projects. But these success stories have rarely been replicated on a scale that has made them cost effective. 'Like expensive boutiques, they are available to the lucky few.' (Binswanger 2000) Thus, a clear and demonstrated commitment of research institutions to development, and a willingness to be held accountable for broader-scale impact, appear not only logical but also a social and economic necessity for future investments in research.

Note

1. Leguminous fallows are natural fallows enriched with planted legumes to improve soil fertility. Biomass transfer is the application of leafy biomass from hedges to crop fields to improve soil fertility.

References

Adesina, A.A., D. Mbila, G.B. Nkamleu, and D. Endamana (2000) 'Econometric analysis of the determinants of adoption of alley farming by farmers in the forest zone of southwest Cameroon', *Agriculture, Ecosystems, and Environment* 80: 255–65

ANAFE (2000) *African Network for Agroforestry Education: ANAFE*, Nairobi: ICRAF

Binswanger, H.P. (2000) 'Scaling up HIV/AIDS programs to national coverage', *Science* 288: 2173–6

Cooper, P.J.M. and G.L. Denning (2000) *Scaling Up the Impact of Agroforestry Research*, Nairobi: ICRAF

Dewees, P.A. and S.J. Scherr (1996) 'Policies and Markets for Non-timber Tree Products', Environmental and Production Technology Division Discussion Paper No. 16, Washington DC: International Food Policy Research Institute

Ericksen, P. and E.C.M. Fernandes (eds) (1998) *Alternatives to Slash-and-Burn Systemwide Programme: Final Report of Phase II*, Nairobi: ICRAF

Franzel, S. (1999) 'Socioeconomic factors affecting the adoption potential of improved tree fallows in Africa', *Agroforestry Systems* 47(1–3): 305–21

Franzel, S., R. Coe, P. Cooper, F. Place, and S.J. Scherr (in press) 'Assessing the adoption potential of agroforestry practices in sub-Saharan Africa', *Agricultural Systems*

Fujisaka, S. (1991) 'Thirteen reasons why farmers do not adopt innovations intended to improve the sustainability of upland agriculture', in *Evaluation for Sustainable Land Management in the Developing World*, IBSRAM Proceedings No. 12(2): 509–22

Grilliches, Z. (1957) 'Hybrid corn: an exploration into the economics of technological change', *Econometrica* 25: 501–23

ICRAF (1997) *ICRAF Medium-Term Plan 1998–2000*, Nairobi: ICRAF

ICRAF (1998) *Building on a Sound Foundation: Achievements, Opportunities and Impact*, Nairobi: ICRAF

ICRAF (2000) *Paths to Prosperity through Agroforestry: ICRAF's Corporate Strategy 2001–2010*, Nairobi: ICRAF

Izac, A.-M. and P.A. Sánchez (in press) 'Towards a natural resource management paradigm for international agriculture: the example of agroforestry research', *Agricultural Systems*

Kwesiga, F.R., S. Franzel, F. Place, D. Phiri, and C.P. Simwanza (1999) '*Sesbania sesban* improved fallows in eastern Zambia: their inception, development and farmer enthusiasm', *Agroforestry Systems* 47(1–3): 49–66

Mercado, Agustin R. Jr., Marcelino Patindol, and Dennis P. Garrity (2001) 'The Landcare experience in the Philippines: technical and institutional innovations for conservation farming', *Development in Practice* 11(3–4): 495-508

Place, F. and P. Dewees (1999) 'Policies and incentives for the adoption of improved fallows', *Agroforestry Systems* 47(1–3): 323–43

Rogers, E.M. and F.F. Shoemaker (1971) *Communication of Innovations*, New York: Free Press

Sánchez, P.A. (1995) 'Science in agroforestry', *Agroforestry Systems* 30: 5–55

Sánchez, P.A. (1999) 'Improved fallows comes of age in the tropics', *Agroforestry Systems* 47(1–3): 3–12

Scherr, S.J. and P.B.H. Hazell (1994) 'Sustainable Agricultural Development Strategies in Fragile Lands', Environmental and Production Technology Division Discussion Paper No. 1, Washington DC: International Food Policy Research Institute

Schutjer, W.A. and M.G. Van Der Veen (1977) 'Economic Constraints on Agricultural Technology Adoption in Developing Nations', USAID Occasional Paper No. 5, Washington DC: USAID

Shah, M. and M. Strong (1999) *Food in the 21st Century: From Science to Sustainable Agriculture*, Washington DC: World Bank

Simons, Anthony J. (1996) 'Delivery of improvement for agroforestry trees', in Mark J. Dieters *et al.* (eds) (1997) *Tree Improvement for Sustainable Tropical Forestry*, Gympie, Australia: Queensland Forestry Research Institute

Participatory design of agroforestry systems: developing farmer participatory research methods in Mexico

Jeremy Haggar, Alejandro Ayala, Blanca Díaz, and Carlos Uc Reyes

Agroforestry systems can take an almost infinite number of different forms, as they have the potential to include any of the crop, animal, and tree species used in agriculture and forestry. This tremendous potential variability allows agroforestry systems to meet the needs of farmers under almost any set of environmental, economic, and social conditions. At the same time, this great plasticity and adaptability of agroforestry makes designing and evaluating agroforestry systems complex (Scherr 1991). Traditional experimentation–validation–dissemination approaches are largely inappropriate for natural resource management innovations such as agroforestry (Rocheleau 1991) because of the long-term nature of tree-based systems and the possibility of multiple solutions. It is not usually appropriate to develop a single production technology for all farmers to apply; rather, it is expected that each farmer will modify any given production technology. Thus a different strategy needs to be developed, incorporating farmers into the research and development process.

Furthermore, rather than trying to homogenise management and treatments, any strategy should exploit the plasticity of agroforestry, by learning from the variations in the way farmers manage agroforestry. Participatory research methods hold the greatest potential for integrating farmers into the process of designing agroforestry systems. On-farm is where the ecological, social, and economic influences that determine the viability of an agroforestry system meet and integrate. Moreover, we believe that farmers are probably the best integrators of these factors.

Opportunities for agroforestry in the Yucatán Peninsula

Although in legal terms, landholding in both indigenous and immigrant communities (*ejidos*) in the southern Yucatán Peninsula is communal, in effect most farmers maintain usufruct rights to between 20 and 100 hectares (ha). The region is primarily covered in secondary semi-evergreen forest, and it receives between 1000 and 1500 millimetres (mm) of rain per year. The soils, derived from limestone, vary greatly depending on topographic position; they include lithosols, rendzinas, luvisols, and vertisols. Farming is based upon shifting cultivation practices that give extremely low yields (0.5–1.0 tonne/ha of maize) plus backyard small-animal production. Crops are supplemented by extracting forest products including timber, *chicle*, honey, and allspice, which may contribute up to half of the household income. Surveys with farmers show their concern to increase the productivity of traditional maize production and diversify production through the planting of fruit and timber trees.

An evaluation of a previous agroforestry project in the region demonstrated farmers' considerable initial and continuing interest in engaging in agroforestry, but a high level of subsequent abandonment of plots by those who undertook it (Snook and Zapata 1998). This suggested that farmers recognised the potential of agroforestry but were experiencing serious problems in implementing it. The principal difficulties they cited were lack of technical support, poor-quality plants, and lack of immediate products. To diagnose the problems in implementing agroforestry, and to determine whether there might exist viable agroforestry systems for the region, we helped farmers to establish eight farmer participatory research groups.

Stages of participatory agroforestry system design

Establishing farmer groups

Farmer research groups were established in two regions of the southern Yucatán Peninsula (see Table 1): Calakmul in the State of Campeche (predominantly mixed-race immigrants from other southern Mexico states), and Zona Maya in the State of Quintana Roo (predominantly indigenous Maya). Only in Campeche were there women members, as in Zona Maya women do not take part in agriculture outside the home garden. In four of the fruit-and-timber groups, farmers were already working with agroforestry. In the other two such groups and in the two improved-fallow groups, researchers suggested the systems and then the farmers opted to collaborate.

The groups varied in their formation: two were based on farmer groups that already existed, three were groups of farmers from a community that had no previous association, and three had no previous association and were composed of farmers from several communities. All participants were self-selecting. Although the research groups that were based on an existing form of association were the quickest to start, internal conflicts related to other activities later affected their functioning. Groups of farmers from the same community without any other formal association between themselves were more successful than groups of farmers from different communities, as there was greater interaction between them outside formal project events. On the other hand, the groups composed of farmers from different communities were able evaluate a technology across a wider range of socio-economic conditions, as was the case with the improved-fallow research groups. Immigrant communities readily adopted the idea of testing new crops and trees. They perceived their experience over the 20 years since they had arrived in the zone as being one of looking for new viable options in a new land – and the options tried had not yet been very successful. Indigenous Mayan communities were

Table 1: Summary of the eight farmer research groups established in southern Yucatán Peninsula, Mexico						
State	Community	No. of farmers		Ethnic group	Production system	Research theme
		Men	Women			
Quintana Roo	Xpichil	8	0	Mayan	Timber and fruit	Associated crops
	Cuauhtemoc	9	0	Mayan and immigrant	Timber and fruit	Tree species trials
	Reforma	7	0	Immigrant	Timber and fruit	Tree species trials
	Zona Maya – four communities	8	0	Mayan	Improved fallows	Establishment methods
Campeche	Calakmul – five communities	8	1	Immigrant	Timber and fruit	Tree species trials
	Narciso Mendoza producers' society	8		Immigrant	Timber and fruit	Legume cover crops
	V. Gomez Farias women's cooperative	1	8	Immigrant	Timber and fruit	Legume cover crops
	Calakmul – three communities	6	3	Immigrant	Improved fallows	Establishment methods

much more reserved about trying new species. Their aims were more to rescue old farming practices in which the younger generation were not interested.

Diagnosis of the potential of agroforestry

Interviews were conducted with all farmers using an open-question semi-structured format based on principal themes. The farmers were asked to present their objectives in working with agroforestry, the problems they had experienced, the solutions they proposed, their future plans in agroforestry, and the limitations they perceived in trying to implement them. This kind of interview, compared with a normal questionnaire, reduced the risk of excluding a key response that concerned the farmers.

Next, the researchers and farmers jointly formulated an agenda of activities during a workshop with each group. First, the results of the diagnosis were presented and reviewed with the farmers. Then researchers and farmers jointly agreed upon the objectives where they had a common interest and the capacity to address. Based on this, both sides proposed actions to resolve the production needs or limitations

Table 2: Farmers' objectives, options tested, problems, and solutions for the Calakmul farmer research group

Farmers' objective	Options in order of preference	Problem	Solution and activity
Produce for home consumption and sale	• Plant staple crops: maize and bean	• Too many weeds	• Test cover crops, researchers provide seed
	• Try new fruit trees: mango, breadfruit, cinnamon, or mamey	• Lack of planting material	• Researchers provide two priority fruit trees
		• Poor growth	• Apply fertiliser
Invest in products for the future	• Plant Spanish cedar and mahogany	• Pests, stem borer that causes poor form	• Training in pest control
Diversify	• Test cash crops: habanero chile, papaya, roselle, or annatto	• Lack of plants/seeds	• Researchers provide seeds for tests
		• Lack of labour	• Community organisation requests financial support from government

identified. These proposals were reviewed and all participants set the priorities. From this an agenda involving activities for research, implementation, and training was developed. Usually both farmers and researchers suggested the activities (see Table 2).

Design and implementation of agroforestry trials

Based on the agenda that emerged from the workshop, one or more trials were developed. Depending on the original objectives, these usually maintained some comparative structure. If the objective was to test different cover legumes for weed control in a fruit-and-timber agroforestry system, then at least one of the treatments would be a control, usually the traditional practice of maize cultivation. In such cases, it was preferable to have some replication, either within or between farms. Nevertheless, the number of replications of any one treatment often varied and reflected the level of interest of the farmers in that option. Where the objective was to test new fruit or crop species, formal controls and comparisons were not thought to be necessary, although any new species was tried by at least two or three farmers.

The farmers implemented the trials on their own and did not receive financial assistance for their labour in establishing and managing them. Researchers covered expenses that implied cash outlay (plants, seeds, agrochemicals), as it was not realistic to expect farmers to make such a high-risk investment. Such inputs, however, were kept to a minimum – that is, they would be within the ability of the farmer to provide if the technology proved successful. Researchers provided the farmers with technical advice both on the management of the experiment and on the crops and trees. The farmers, however, made their own decisions on how to manage the system.

Evaluation of agroforestry trials

Evaluation included criteria that were important to farmers as well as those which concerned the researchers. Farmers and researchers made the field evaluations jointly, and the researchers presented all of the data collected to the farmers. Many of the criteria the farmers evaluated, such as taste of product, were not readily quantifiable but were critical to the acceptability of an option. Workshops were conducted in which farmers ranked or scored different options as a group (Ashby 1990). Farmer groups evaluated component species for agroforestry systems and then noted any factors (modifiers) that might limit the potential of the species (see Table 3).

Table 3: Priority of components for an agroforestry system by the farmers of the Cuauhtemoc research group, by scoring relative importance of the different components (1 = very important, 2 = moderately important, 3 = of lesser importance) and overall farmer preference (1 is most preferred)

| | Scoring of importance | | | |
	Home consumption	Sale	Modifying comments	Score on overall farmer preference
Annual crops				
Jamaica	3	1		1
Sesame	3	2		2
Maize	1	3		3
Beans	1	3		3
Perennial crops				
Plantain	1	1	Only deep soil	1
Annatto	3	2		2
Pineapple	3	2	Only deep soil	3
Cassava	2	3	Only deep soil	4
Fruit trees				
Avocado	1	2		1
Mamey	1	2		1
Mango	1	2		1
Sapotillo	1	2		1
Star apple	2	3		2
Soursop	2	1	Fruit rot	2
Tamarind	1	3		2
Sweetsop	2	3		3
Nance	3	2		3
Custard apple	2	3		4
Cashew	3	3		5
Timber trees				
Spanish cedar		1	Fastest growth	1
Mahogany		1		2
Ciricote		2		3

All components tested by some or all of the farmers

Fundamental to the joint farmer–researcher evaluation was an integrated evaluation of the trials themselves. Different quantitative and qualitative evaluations were integrated by forming a matrix of the ranked qualifications. For example, the Narciso Mendoza group ranked the cover legumes that were tested in a fruit-timber agroforestry system according to the services provided (weed control and mulch production), yield, and quality of product (see Table 4). Rather than look for a single 'best result', these qualifications were used to identify different production strategies that would be adapted to the different objectives of the farmers. In this case, the best options for food production were varieties of cowpea, while the best for weed control was canavalia or mucuna (Haggar and Uc Reyes 2000).

Adaptation of participatory methods to different circumstances

Most of the farmers participating in testing fruit-timber tree agro-forestry systems had some prior experience with this system, so it was

Table 4: Farmers' evaluation of cover legumes in a fruit–timber agroforestry system by the Narciso Mendoza farmer group (5 is high, 1 is low)

	Services*		Yield†		Quality of product‡	
	Rank in group	Rank of best of each group	Rank in group	Rank of best of each group	Rank in group	Rank of best of each group
Bush bean		2		1		3
Cowpea (var. Xpelon)	5		4		5	
Cowpea (var. Andalon)	5		5		3	
Black bean (var. Jamapa)	3		3		4	
Red bean (var. Michigan)	1		1		1	
Red bean (var. Flor de Mayo)	1		1		1	
Cover legumes		3		3		1
Mucuna	3		2		3	
Canavalia	3		3		2	
Lima bean	1		1		1	
Other legumes		1		2		2
Soya	1		2		5	
Peanut	3		4		3	
Cowpea (var. Lentejito)	4		5		4	
Pigeon pea	5		4		2	
Clitoria	2		1		1	

* Services of weed control, mulch production, with modifying comments on disease susceptibility and competition with trees
† Yield based on data taken by the farmers
‡ Quality of product for human or animal consumption

possible for them to diagnose their problems. But improved fallows were a totally new concept, and farmers were unfamiliar with cultivating the species – *mucuna* and *leucaena*. It was therefore necessary for researchers to design the initial trial with those farmers who were interested in these fallows in a way that farmers could later modify as they gained experience with the plants and the system. To initiate the process and demonstrate the idea, they presented the farmers with two highly contrasting improved fallows. One improved fallow was planted with *leucaena*, a shrub, and the other with *mucuna*, an herbaceous leguminous climber. After two years of establishing improved fallows with these species, farmers identified a technique for each. To establish *leucaena*, they preferred to broadcast large quantities of seed before burning the plot. Rather than sowing *mucuna* for an improved fallow, they preferred to use it in the traditional method as a green manure within or between maize crops. Thus, after gaining experience with the system these farmers could redesign and adapt the original system to meet their own conditions and needs.

Impact of participatory research and the empowerment of farmers

There has been some concern that participatory research methods may create only local solutions for local problems. Obviously, it is not possible to assist every community to have its own participatory research group. To ensure that participatory research provides solutions for more than just those individuals who directly take part in it, both the communities and the participants within the communities should be selected to represent the range of ecological, social, and economic conditions over which an impact is expected. It must be recognised, however, that investing in research may be beyond the capacity of the poorest farmers.

Aside from the technological recommendations *per se*, the greatest impact of participatory research arises from its emphasis on empowering farmers to act in the research and development process. Farmers' trials were used as demonstration plots to disseminate the results of the research to other farmers and to other communities. Farmer experimenters themselves promoted the results of their experience.

In the future it is hoped that the farmer research groups will develop greater independence with more limited external facilitation of their activities, similar to the local agricultural research councils (CIALs) widely implemented in Central and South America (Ashby and Sperling 1995). However, because of the complexity of agroforestry systems and the long-term investment necessary to produce trees, a longer-term partnership between researchers and farmers than is normally undertaken may be desirable, to establish a CIAL. All the communities we work with belong to a community organisation, either the Xpujil Regional Council (CRASX) or the Zona Maya Organisation of Forest Producers (OEPFZM). These provide a forum where the farmers present the results of their research to the leaders of the organisations. They are using the results of participatory research to adapt government development projects to better meet the needs of their members. One OEPFZM now has a fruit-and-timber agroforestry project that is working with 200 farmers. Government development projects in both Quintana Roo and Campeche are using both the fruit-

and-timber and the improved-fallow work of the farmers to teach extension workers how to provide farmers with alternatives to slash-and-burn agriculture.

Acknowledgements

We would like to acknowledge the following people who have collaborated over the past three years in developing participatory research methods for agroforestry: Simon Anderson, Bernadette Keane, Julieta Moguel, Sabine Gundel, and Ramon Camara of the Diagnosis and Participatory Research Group, University of Yucatán; Bernard Triomphe of the Rockefeller Foundation; and not least the farmers of Calakmul and Zona Maya. We also thank John Weber, Peter Cooper, and Steve Franzel for their valuable comments on this paper; and the Ford Foundation, which has supported this research for three years.

Refcrences

Ashby, J.A. (1990) *Evaluating Technology with Farmers: A Handbook*, CIAT Publication No. 187, Catie, Colombia: Centro Internacional de Agricultura Tropical

Ashby, J.A. and L. Sperling (1995) 'Institutionalizing participatory client-driven research and technology development in agriculture', *Development and Change* 26: 753–70

Haggar, J.P. and C. Uc Reyes (2000) 'Investigación participativa en el uso de leguminosas de cobertura en sistemas agroforestales en Calakmul, Campeche', *Agroforesteria en las Americas* 7(28): 16–20

Rocheleau, D.E. (1991) 'Participatory research in agroforestry: learning from experience and expanding our repertoire', *Agroforestry Systems* 15: 111–38

Scherr, S.J. (1991) 'On-farm research: the challenge of agroforestry', *Agroforestry Systems* 15: 95–110

Snook, A. and G. Zapata (1998) 'Tree cultivation in Calakmul, Mexico: alternatives for reforestation', *Agroforestry Today* 10: 15–18

Participatory domestication of agroforestry trees: an example from the Peruvian Amazon

John C. Weber, Carmen Sotelo Montes, Héctor Vidaurre, Ian K. Dawson, and Anthony J. Simons

Farming communities in the Peruvian Amazon depend upon more than 250 tree species for construction material, fence posts, energy, fibres, resins, fruits, medicines, and service functions such as soil conservation and shade (Sotelo Montes and Weber 1997). However, because of deforestation, forest fragmentation, and overlogging, the diversity and the quality of valued tree species are declining around many rural communities. The resultant loss to local communities in income, self-reliance, and nutritional security is often severe. In addition, national and global environmental benefits of forests are reduced.

Most farmers practise traditional slash-and-burn agriculture in the Peruvian Amazon (Denevan and Padoch 1988). Because the typically acidic soils lack sufficient nutrients for sustainable, repeated harvests of annual crops, farmers cut and burn the forest to release accumulated nutrients in the woody biomass. This allows one to three years of cropping (rice, maize, cassava) before the fields are left to fallow or are converted to pasture. A 20-year forest fallow is considered necessary to restore soil fertility for a sustainable three-year cropping phase (J.C. Alegre, personal communication). Since an average farm is only 30 hectares (ha), and 2–3 ha are typically cleared annually for crop production, a 20-year rotation is not an option for most farmers. So most farmers decrease the fallow period to five years or less, resulting in degraded soils and low crop yields.

Slash-and-burn agriculture fragments and alters the forest habitat, resulting in poor natural regeneration of many valuable tree species. In addition, farmers and loggers cut the best timber trees in their forests, without leaving high-quality trees to produce seed for natural regeneration (Weber et al. 1997). In time, farmers no longer have access to high-quality tree seed for their agroforestry systems. When crop yield and forest value decrease, farmers often migrate to open up more land, thereby repeating the cycle of deforestation, forest fragmentation, soil degradation, and poverty.

The International Centre for Research in Agroforestry (ICRAF) and its partners are working to counter this cycle and to help ensure that farming communities and the global community continue to derive the benefits that trees provide. In this paper, we describe a tree domestication project underway with farming communities in the Peruvian Amazon. Domestication is defined as an iterative process involving the identification, production, management, and adoption of desirable tree germplasm. The project aims to provide increased productivity and long-term sustainability for farm forests, while also empowering farming communities to conserve tree genetic resources.

Principles of farmer-driven tree domestication

Farmers domesticate trees by bringing them into cultivation, adapting them to their needs and environmental conditions by deliberately or inadvertently selecting for certain characteristics, and applying particular management strategies (Leakey and Simons 1998). To develop and implement an effective domestication strategy, farmers and researchers should collaborate from the outset. This is because farmers, who are the principal beneficiaries of tree domestication, can best identify their needs in a research programme. Farmers also have valuable knowledge about tree species that can guide the research programme.

Identifying farmers' preferences for agroforestry trees is the first step in participatory tree domestication. Following guidelines developed for priority setting (Franzel *et al.* 1996), we conducted farmer-preference surveys and solicited advice from experts in forest products, markets, and other disciplines. We learned that farmers would like to cultivate more than 150 tree species, and we identified 23 of these as high priority for domestication (Sotelo Montes and Weber 1997). Domestication projects have begun for four of the species that figure significantly in the farm economy (Labarta and Weber 1998): *Bactris gasipaes* Kunth, *Calycophyllum spruceanum* Benth., *Guazuma crinita* Mart., and *Inga edulis* Mart.

Documenting farmers' knowledge about variation within a tree species is an essential component of participatory tree domestication. The documentation provides testable hypotheses for research that can accelerate the delivery of improved tree planting material to farmers.

For example, *Bactris gasipaes* (peach palm) is an under-utilised food crop that many farmers cultivate on a small scale. Some farmers prefer a starchy fruit for flour, while others prefer an oily fruit for cooking oil. Farmers have learned to use visual identification to tell which palms are better for these products. In their experience, fruits with red, waxy coats have higher oil content than fruits with red or yellow, non-waxy coats. This hypothesis about variation in fruit traits is being tested experimentally, and if proved correct, will allow farmers and researchers to select the best genetic material for multiplication quickly and inexpensively.

When documenting farmers' knowledge and perceptions about variation within tree species, it is essential to recognise potential gender differences, because men and women may value different tree species. For example, many women cultivate trees in home gardens for fruit, medicines, and other products that they use in the household and sell in local markets. In one case study (Potters 1997), women recognised six varieties of *Inga edulis*, which they would cultivate specifically for fruit, shade, or firewood. They distinguished the varieties based on pod size and on the size, shape, and colour of the leaves. According to their experience, certain varieties have tastier fruits, while others are better for shade. The women also perceived a correlation between seed colour and fruit production. In their experience, trees that develop from black seeds produce many fruits, but trees from yellow seeds do not produce much fruit. Men, on the other hand, dedicate considerable time to land clearing, charcoal production, house construction, and fence building. With experience, they learn which species are best for different uses, and recognise variation within some of their most valued species. For example, they cultivate *C. spruceanum* for timber and charcoal in secondary forests. In their experience, *C. spruceanum* trees with dark reddish-brown bark and few knots on the stem have dense wood and are best for sawn timber, whereas the wood of trees with light-coloured bark is not dense, is easy to split with an axe, and is best for firewood and charcoal.

Selecting improved tree planting material with farmers

Through appropriate selection strategies, farmers can achieve improvement in timber-tree form, fruit quality, and other commercially important traits (Simons *et al.* 1994). Most tree species include considerable genetic variation, which provides opportunity for selection and improvement. The challenge is to determine efficient methods for characterising variation and for selecting the best genetic material with farmers for agroforestry systems that are often complex and vary from farm to farm. Non-traditional approaches, which involve farmers as collaborators throughout the research process, are generally required.

An example of a non-traditional approach to improvement is illustrated by on-farm provenance trials of two timber-tree species, *Calycophyllum spruceanum* and *Guazuma crinita*. The former is also valued for firewood, charcoal, and construction poles. Both species are relatively fast growing, and can be harvested in 3–20 years, depending on the product desired. Researchers and farming communities over a large geographic area identified 11 extensive natural populations (provenances) of each species in the Peruvian Amazon and then collected seed from these populations in 1996. Trees were not selected based upon their physical appearance; they were sampled following a 'systematic collection' strategy (35 trees collected at random in each population, with a minimum distance of 100 m between trees) designed to ensure a representative sample of the variation within natural populations (Dawson and Were 1997). The seed was used to establish on-farm provenance trials in the Aguaytía watershed in 1998. The principal objective of the ongoing trials is to identify the most promising provenances for different products under various rainfall and soil conditions in the watershed. The Aguaytía is representative of many watersheds in the western Amazon basin; in general, soils are more fertile and rainfall is higher in the upper portion of the watershed. The study area extends over a distance of approximately 80 km. Along this 80 km, elevation increases from approximately 180 to 300 m, and annual rainfall increases from approximately 1800 to 3500 mm. Temperature data are not available, but average annual temperature is approximately 26°C.

Farmers were selected in the lower, middle, and upper parts of the watershed to participate in the on-farm provenance trials. We selected 20 farmers based upon their interest in the project, experience in tree management, and standing in the community as innovators and leaders. Each farmer has one replication of the trial. The farmers participate actively in evaluating tree growth, and they provide useful information about selection criteria, such as the hypothesis mentioned above concerning bark and wood characteristics of *C. spruceanum*. If results from the provenance trial prove this hypothesis correct, farmers will be able to select rapidly the best trees for sawn timber or for charcoal, and multiply the seed for personal use and sale.

Preliminary results of the trials illustrate potential gains that farmers can realise from selection. After one year in the field, the local provenance of *G. crinita* from the Aguaytía watershed was significantly taller than the other provenances, suggesting that seed from the Aguaytía watershed would be best for reforestation in that watershed and in other watersheds with similar environmental conditions (Sotelo Montes *et al.* 2000). However, wood density and other traits should also be evaluated before recommending the best seed source. In *C. spruceanum*, for example, wood density varies greatly among provenances and environments, and no single provenance performs best in all environments (see Figure 1: there were only eight provenances in the trial). Some provenances have higher wood density in the upper watershed, with more favourable growing conditions, while others perform better in the middle and lower watershed where soils are less fertile and drier. This indicates that different provenances of *C. spruceanum* are likely to be better adapted to different parts of the watershed.

An ideal provenance for timber and energy would grow rapidly and produce wood of high density and high heat content. Identifying this provenance requires evaluating growth and wood traits together. We illustrate this with *C. spruceanum*, using a statistical technique that summarises variation in growth and wood traits into component 'traits' (principal component analysis). Provenance means for two component traits are plotted; the horizontal axis summarises branch-wood density and heat content, while the vertical axis summarises growth, which

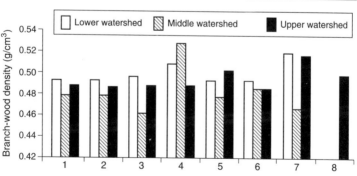

Figure 1: Variation in branch-wood density at 18 months among provenances of *Calycophyllum spruceanum* tested in the Aguaytía watershed, Peru

Provenance: 1 Lagunas, 2 Barranca, 3 Pastaza, 4 Jenaro Herrera, 5 Tamshiyacu, 6 Mazan, 7 Pevas, 8 Von Humboldt (tested only in upper watershed)

includes tree height, stem diameter, number of stem nodes and branches. In this case, the best provenances fall in the upper right quadrant of the plot. In the lower watershed, one provenance showed a 'win–win' response with the best growth and wood characteristics (see Figure 2). In the upper watershed, several provenances expressed these characteristics and could be considered as seed sources for reforestation (see Figure 3).

Figure 2: Plot of component traits of *Calycophyllum spruceanum* provenances tested in the lower Aguaytía watershed, Peru

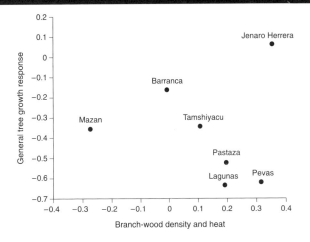

Figure 3: Plot of component traits of *Calycophyllum spruceanum* provenances tested in the upper Aguaytía watershed, Peru

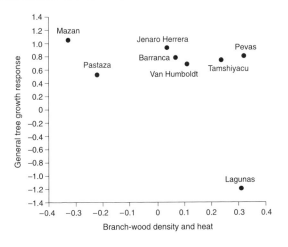

Improved performance is clearly important when selecting tree populations for cultivation, but we must also consider genetic diversity. Diversity is needed to enhance the capacity of planting material to adapt to changing user requirements and environmental conditions (Simons *et al.* 1994). Moreover, most tree species do not tolerate inbreeding; if inbreeding during cultivation significantly reduces their genetic diversity, the next generation will be less vigorous and less productive.

Tree domestication practices can have a conservation function if they ensure that planted material is productive, adapted, and genetically diverse. Provenance and progeny trials are useful for selecting more productive and adapted tree planting material for different environmental conditions, but sophisticated molecular techniques are required to assess overall genetic diversity. Using these techniques, we identified the most diverse provenances of *C. spruceanum* (Russell *et al.* 1999) and recommended that farming communities manage them for *in situ* conservation and seed production (Sotelo Montes *et al.* 2000). In *Inga edulis*, we are concerned that domestication has reduced genetic diversity on farms, compared with wild populations. Preliminary results confirm this hypothesis (T. Pennington, personal communication) and warrant introducing germplasm to increase the genetic diversity on farm and the participatory management methods needed to maintain the diversity.

Demonstrating the risk of poor tree adaptation to farmers

In most tropical countries, few mechanisms control the source of seed used for tree-planting projects. Seed may be collected in one region and used in another with different environmental conditions, without any knowledge of its potential adaptation. Seed zones and transfer guidelines based upon ecological and sometimes genetic criteria have been proposed as ways to minimise the risks of disappointing tree performance. Normally, it is assumed that transferring seed from one environment to another imposes some degree of risk and that seed from local sources is generally better adapted to local environmental conditions than seed from foreign sources.

To test questions of adaptation, it is important to select an appropriate 'field laboratory' so that results can be extrapolated over larger areas. As mentioned, the Aguaytía watershed is representative of many others in the Peruvian Amazon. In 1998, farmers selected 200 trees each of *C. spruceanum* and *G. crinita* that had been sampled throughout the lower, middle, and upper parts of the Aguaytía watershed. These were 'targeted collections' (Dawson and Were 1997), in which farmers selected their best trees based upon desirable timber characteristics (straight bole without bifurcation, few knots, thin branches, relatively small canopy). Progeny of the 200 trees were established in progeny trials on 15 farms in the lower, middle, and upper parts of the same watershed in 2000-1. The results of these trials will allow us to quantify the effect of a hypothetical seed transfer from one part of the watershed to another. For example, we hypothesise that progeny of trees collected in the upper part of the watershed will grow best in the replications located in the upper watershed and least well in the replications located in the lower watershed. The relative differences in progeny performance in the lower, middle, and upper watershed will allow us to estimate potential production losses associated with a given seed transfer from one part of the watershed to another. Farmers and development workers will then be able to decide if the risk is acceptable, when balanced against other factors. Conclusions drawn from the trials can also be extrapolated to other watersheds with similar environmental conditions.

Accelerating the delivery of high-quality planting material to farmers

Once appropriate planting material has been identified, what is the most efficient way to produce and disseminate it? Farmers consistently cite the lack of high-quality tree seed as a major constraint to diversifying and expanding their agroforestry practices (Simons 1996). Providing farmers with high-quality tree seed in a timely manner, therefore, is one of the principal objectives of tree domestication. A traditional approach to tree improvement involves a number of sequential steps – species selection trials, provenance trials to identify the best seed sources of a species, progeny tests to identify the best mother trees within a selected seed source, collection of germplasm from the best mother trees to establish seedling or clonal seed orchards, and, finally, the production of high-quality seed from orchards for dissemination to users. This process is too time consuming and expensive to undertake for most tree species, and the work, if not carefully planned, may seriously reduce the genetic diversity in tree-planting material. Furthermore, the formal sector of governmental and non-governmental organisations that produce and disseminate tree seeds cannot meet the growing demand for quality planting material.

An alternative approach is to involve key farmers not only in selecting the most promising planting material, but also in multiplying and disseminating it. This approach reduces the number of steps to the end user, while measures are taken to maintain genetic diversity and quality. For example, on-farm provenance or progeny trials of *C. spruceanum* and *G. crinita* can be converted into seed orchards after approximately three years, thereby dramatically shortening the time required to deliver selected planting material to farmers. Farmers and researchers will thus be able to select the best trees to retain for seed production and can then cut the other trees in each replication. Farmers can use or sell the cut trees for construction poles, firewood, and other products, and manage stump sprouts for future harvests. The orchards themselves can satisfy the entire demand for planting material in the watershed and have a decentralised, *circa situ* conservation function on farm. Participating farmers are being organised into networks to produce and sell high-quality seed, seedlings, and timber to organisations involved in tree planting projects and to timber companies. This will be a new form of small-business enterprise in Peru. The earnings will depend on the scale of production, but it is estimated that farmers will be able to earn at least US$1000 per year from the seeds, seedlings, and timber in a 1 ha lot of *C. spruceanum* or *G. crinita*.

Adoption of the tree domestication methodology

National and local institutions, non-governmental organisations, and private enterprise in Peru are becoming increasingly aware of the need to conserve tree genetic resources and manage them sustainably. We are working with several institutions and selected timber companies to promote the use of improved seed and certified timber production for the international market. We are also working with the National Institute for Natural Resources (INRENA) to institutionalise tree domestication in the Peruvian Amazon. Through INRENA's extensive network for reforestation, it will be possible to reach thousands of farmers in the next five years. Through INRENA's lobbying efforts in the Peruvian government, policy changes are being introduced in the forestry laws to promote the sustainable use and conservation of tree genetic resources for future generations.

Lessons learned and conclusions

Farmers are frustrated with the low prices for traditional food crops and the unstable prices for perennial crops like coffee and cocoa. From past experience, they are sceptical of new 'boom' crops like palm hearts for the gourmet market. They are eager for alternatives in which they have a comparative advantage in production, and many now see tree crops for timber, fruit, energy, medicine, and seed as a good investment.

Farmers should have a strong vested interest in conserving tree genetic resources, since they are the first to suffer if these resources decline in value. But they will become encouraged to conserve only if they see tangible economic benefits. The challenge, therefore, is to engage key farmers in tree domestication research, quantify the economic benefits, and use the farms to demonstrate the economic potential to others in the community. This is not easy, because most farmers in the region lack a 'tree-planting culture' and think in the very short term. Getting most farmers to think ahead for a longer term may require generations. It may ultimately depend on the success of the national educational system and non-governmental conservation organisations.

There is a clear need to use genetic resources sustainably and conserve valuable tree species for the future economic development of resource-poor farmers in the Peruvian Amazon. We believe that the best way to achieve this goal is by promoting participatory tree domestication and conservation-through-use, where farmers themselves manage the resources, with technical assistance from international and national institutions (O'Neill *et al.* in press).

We cannot scale this project up to the national level with our limited resources. But we can train and motivate national and local institutions and private enterprise to adopt these methods and scale up tree domestication. The major challenge is to demonstrate the short, medium, and long-term economic potential that can be realised by domesticating trees and conserving tree genetic resources through wise and careful use of them.

References

Dawson, Ian K. and James Were (1997) 'Collecting germplasm from trees: some guidelines', *Agroforestry Today* 9(2): 6–9

Denevan, William M. and Christine Padoch (eds) (1988) *Swidden-Fallow Agroforestry in the Peruvian Amazon*, Advances in Economic Botany Series, Volume 5, New York: New York Botanical Garden

Franzel, Steve, Hannah Jaenicke, and Willem Janssen (1996) *Choosing the Right Trees: Setting Priorities for Multipurpose Tree Improvement*, The Hague: International Service for National Agricultural Research

Labarta Chávarri, Ricardo and John C. Weber (1998) 'Valorización económica de bienes tangibles de cinco especies arbóreas agroforestales en la Cuenca Amazónica Peruana', *Revista Forestal Centroamericana* 23: 12–21

Leakey, Roger R.B. and Anthony J. Simons (1998) 'The domestication and commercialization of indigenous trees in agroforestry for the alleviation of poverty', *Agroforestry Systems* 38: 165–176

O'Neill, Gregory A., Ian K. Dawson, Carmen Sotelo Montes, Luigi Guarino, Dean Current, Manuel Guariguata, and John C. Weber (in press) 'Strategies for genetic conservation of trees in the Peruvian Amazon basin', *Biodiversity and Conservation*

Potters, Jorieke (1997) 'Farmers' Knowledge and Perceptions about Tree Use and Management: The Case of Trancayacu, a Peruvian Community in the Amazon', MSc thesis, Wageningen, The Netherlands: Wageningen Agricultural University

Russell, Joanne R., John C. Weber, Allan Booth, Wayne Powell, Carmen Sotelo Montes, and Ian K. Dawson (1999) 'Genetic variation of *Calycophyllum spruceanum* in the Peruvian Amazon basin, revealed by amplified fragment length polymorphism (AFLP) analysis', *Molecular Ecology* 8: 199–204

Simons, Anthony J. (1996) 'Delivery of improvement for agroforestry trees', in Mark J. Dieters, Colin A. Matheson, Garth D. Nikles, Chris E. Harwood, and Steve M. Walker (eds) *Tree Improvement for Sustainable Tropical Forestry*, Gympie, Australia: Queensland Forestry Research Institute

Simons, Anthony J., Duncan J. MacQueen, and Janet L. Stewart (1994) 'Strategic concepts in the domestication of non-industrial trees', in Roger R. B. Leakey and Adrian C. Newton (eds) *Tropical Trees: The Potential for Domestication and the Rebuilding of Forest Resources*, London: HMSO

Sotelo Montes, Carmen and John C. Weber (1997) 'Priorización de especies arbóreas para sistemas agroforestales en la selva baja del Perú', *Agroforestería en las Américas* 4(14): 12–17

Sotelo Montes, Carmen, Héctor Vidaurre, John C. Weber, Anthony J. Simons, and Ian K. Dawson (2000) 'Producción de semillas a partir de la domesticación participativa de árboles agroforestales en la amazonía peruana', in Roberto Salazar (ed.) *Memorias del Segundo Symposio sobre Avances en la Producción de Semillas Forestales en América Latina*, Turrialba, Costa Rica: Centro de Agricultura Tropical y de Enseñanza

Weber, John C., Ricardo Labarta Chávarri, Carmen Sotelo Montes, Angus W. Brodie, Elizabeth Cromwell, Kate Schreckenberg, and Anthony J. Simons (1997) 'Farmers' use and management of tree germplasm: case studies from the Peruvian Amazon Basin', in Anthony J. Simons, Roeland Kindt, and Frank Place (eds) *Proceedings of an International Workshop on Policy Aspects of Tree Germplasm Demand and Supply*, Nairobi: International Centre for Research in Agroforestry

Facilitating the wider use of agroforestry for development in Southern Africa

Andreas Böhringer

Southern Africa is one of six eco-regions in which the International Centre for Research in Agroforestry (ICRAF) currently operates. Work in this region started in Malawi, Tanzania, Zambia, and Zimbabwe between 1986 and 1987 with efforts to diagnose farming-system constraints and to design agroforestry interventions. These countries are located within the savannah woodland eco-zone, or *miombo*, which is characterised by one rainy season, receiving 700–1400 mm of rainfall annually and a long dry season from May to October. An upland plateau ranging in altitude from 600 to 1200 m dominates the region. The soils, predominantly ferralsols, acrisols, and nitosols according to FAO classifications, are generally poor in nutrients. Mixed farming is dominant in the region, with crops and livestock being integrated only very loosely in traditional systems. The degree of cropping and livestock keeping varies depending according to ethnic background and the availability of natural grazing land, the latter having declined in recent decades because of population growth. Maize is the most important staple crop throughout the region, and its production and prices are often dictated more by politics than economics, especially as the urban electorate becomes increasingly influential. Cassava, sweet potato, sorghum, millet, and various grain legumes are other important subsistence crops. Food insecurity is common in the region, and is underscored by a several-fold increase in maize imports in most countries over the past decade. Access to safe drinking water, basic health services, and markets is critical in most rural areas.

Key farming-system constraints that have been identified are all associated with widespread and advancing degradation of the natural resource base and accelerated deforestation. Caused principally by increasing human populations, both have led to a widespread decline in soil fertility, increased soil erosion, and shortages of fuelwood and seasonal fodder, to mention only the most severe effects felt on-farm. Continuous cultivation of maize exacerbates the depletion of soil fertility, with nitrogen being particularly critical for good production in most parts of the region. Research started in 1987 to develop agroforestry technologies to address these problems. Project sites were gradually established in Shinyanga and Tabora in Tanzania, Zomba in Malawi, Chipata in Zambia, and Harare in Zimbabwe. Since 1997, ICRAF has been breaking new ground as a research centre in Southern Africa by getting more proactively involved in development. This engagement seeks primarily to accelerate the impact of agroforestry in the region.

This paper reports on the process and outcomes of research-driven technology development and how it has recently evolved into a more client-driven process. This shift looks promising as a way to reach large numbers of particularly poor households, a disadvantaged group that is of top priority to ICRAF. Agroforestry technologies that are now available have great potential to improve the livelihoods of many in the region. This paper first assesses development trends in Southern Africa and describes agroforestry options addressing farmers' constraints. The problems and successes ICRAF has experienced in facilitating the wider use of agroforestry in the region are a further focus, so that lessons learned can be shared with a wider audience. The paper highlights the role of agroforestry as a learning tool in helping local communities to become more capable of developing technological and other kinds of innovations.

Development trends in Southern Africa and agroforestry opportunities

For the purposes of this paper, the Southern African region is the area similar to that covered by the original Southern Africa Development Community (SADC), with 11 member countries: Angola, Botswana, Lesotho, Malawi, Mozambique, Namibia, South Africa, Swaziland, Tanzania, Zambia, and Zimbabwe. These countries cover an area of 6,823,490 km² and have a total population of slightly over 142 million. Only small proportions of land are classified as arable, ranging from as little as 1 per cent for Namibia to a maximum of 18 per cent for Malawi. Populations are therefore exerting considerable pressure on available arable land, with maximum densities in rural areas reaching as many as 350 persons per km² in southern parts of Malawi. Population growth rates have fallen well below 3 per cent recently (see Table 1), mostly an effect of the AIDS pandemic, but are still large enough to trigger further significant population increases over the next 20 years. Some key development indicators of the five countries where ICRAF currently operates in Southern Africa are summarised in Table 1. Assuming for Mozambique, where no data are available, that 60 per cent of the total population is below the poverty line, this would mean that 44 million people currently live in absolute poverty in these five countries. If the average household size across the region is six persons, this would translate into an approximate total of 14.67 million poor households.

The overall goal in widening the use of agroforestry in the region is to make an impact on people's livelihoods, in particular on food security and poverty, and to reverse the progressive degradation of the natural resource base. Since socio-political, economic, and environmental conditions govern any large-scale use of agroforestry, and these also change constantly, it is important to predict likely future trends so that the best agroforestry interventions can be identified and innovations developed together with farmers in good time. This was done in a regional strategic planning exercise that ICRAF facilitated in early 1999, where major institutional stakeholders were represented.

Table 1: Some development indicators for selected countries in Southern Africa (compiled from <http://www.odci.gov/cia>)

Indicator	Tanzania	Zambia	Malawi	Mozambique	Zimbabwe
Total population (m)*	31.27	9.66	10.00	19.12	11.16
Population growth rate (%)*	2.14	2.12	1.57	2.54	1.02
Life expectancy at birth (yrs)*	46.17	36.96	36.30	45.89	38.86
Literacy (% total pop.)†	67.8	78.2	56.4	40.1	85.0
GDP per capita (US$)‡	730	880	940	900	2400
Contribution of agriculture to total GDP (%)§	56	23	45	35	28
Population below poverty line (%)¶	51.1	86.0	54.0	No data	25.5
UNDP human poverty index (%) (rank)ǁ	29.8 (54)	38.4 (64)	42.2 (72)	49.5 (79)	29.2 (53)

* 1999 estimates
† Aged 15 years and over who can read and write, 1995 estimates
‡ Purchasing power parity, 1998 estimates
§ 1995 estimates for Malawi, 1996 estimates for Tanzania and Mozambique, 1997 estimates for Zambia and Zimbabwe
¶ 1991 estimates for Tanzania, Malawi, and Zimbabwe, 1993 estimates for Zambia
ǁ 1997 data published in the UNDP Human Development Report 1999

Table 2: Predicted future development trends in Southern Africa and possible opportunities for agroforestry

Development trends	Opportunities for agroforestry
Socio-political	
• Disintegration of the extended family and loss of traditional values	• More land becoming available in areas of urban drift
• Wider access to information	• Peri-urban and urban agroforestry production for niche markets
• Increased urbanisation	
• Declining social status of farming	• Empowerment of grassroots institutions that drive the scaling up of agroforestry use
• Decentralisation of decision making and advancing democratisation	
• Change of demographic structure and declining overall productivity in communities because of AIDS	• Increased accountability of local institutions
• Regional political integration	• Easier adoption of gender- and age-neutral agroforestry technologies
Economic	
• Increased poverty and widening of gap between rich and poor	• Increased private investment into processing and marketing of agroforestry products
• Decline of real incomes and continued devaluation of local currencies	• Emerging cottage industries and adding of value to products at local level
• Continued dependency on external aid	
• Increased privatisation	• Growing markets for 'green' products in urban areas
• Regional economic integration with South Africa emerging as dominant player	• Increased demand for substitution of expensive external agricultural inputs such as fertiliser and feed
Environmental	
• Increased deforestation	• Stabilisation of overall natural resource base through agroforestry
• Decline in biodiversity	
• Soil and water degradation	• Conservation of economically important indigenous trees through domestication and marketing
• More pronounced fluctuations in seasonal rainfall (droughts and floods)	
	• Stabilisation of land-use systems through diversification and ecological buffering (trees as risk buffers)

Table 2 summarises the results of this workshop and highlights future opportunities for agroforestry in the region.

Developing agroforestry technology options

ICRAF's research agenda in Southern Africa focuses on replenishing soil fertility, producing fuelwood and fodder, and evaluating suitable tree germplasm, including the domestication of fruit trees indigenous to the *miombo* woodland. The research effort first diagnosed farmers' priority problems, then followed up with on-station and on-farm research. By 1997, approximately 5000 farmers were participating in on-farm research in the four countries where ICRAF has project offices.

The main agroforestry technologies developed were improved fallows, mixing of coppicing trees with crops, annual relay cropping of trees, fodder banks, rotational woodlots, and planting of indigenous fruit trees. These options, now used by large numbers of farming families, are described briefly below.

Improved fallows

A piece of land is dedicated to fallowing with nitrogen-fixing tree species for a minimum of two growing seasons. During at least one season, trees take up the entire field and no crops can be planted. The tree most successfully used is *Sesbania sesban*, but farmers also plant *Tephrosia vogelii,* both species being native to Africa. The main objective is to achieve household food security in staple foods, most importantly maize, in situations where land availability is not severely limited (population densities <60 persons/km^2).

The technology aims to replenish soil fertility, in particular nitrogen, with little or no external inputs such as fertilisers, resulting again in significantly increased maize yields after two years of fallowing. Farmers have intensified improved fallows by intercropping during the first year while the trees are being established. The main requirements of the technology are the availability of land, high demand for labour, availability of enough water to establish the trees successfully, and the need to protect the improved fallows during the dry season from fires and free-ranging livestock (the latter being less of a problem for *Tephrosia*). When large areas are planted to one tree species, insect pests may become a problem; this is already occurring in some places with the *Mesoplatys* beetle on *Sesbania*. (For a more detailed description of improved fallows using *Sesbania*, see Kwesiga and Beniest 1998.)

Mixing coppicing trees and crops

Nitrogen-fixing trees that can tolerate continuous cutting back, such as *Gliricidia sepium* from Central America, are mixed in and grown with crops in the field. Trees are coppiced in such way that they do not interfere with the crop, yet large amounts of cut biomass can be recycled as green manure over many years. The main objective is to achieve household food security in staple foods in situations where the availability of land is severely limited, such as in parts of southern Malawi where population densities are over 100 persons/km^2.

The technology aims at replenishing soil fertility, in particular nitrogen, with little or no external inputs such as fertilisers being required, resulting again in significantly increased maize yields, usually three years after tree planting. After this time, and provided that the trees are managed rigorously, the technology has been shown to perform well for at least eight years without any need for fallowing the land.

One limitation is that a three-year waiting period is needed before trees reach their full biomass productivity and benefits become tangible. Furthermore, a considerable amount of labour may be required for tree management, but farmers appreciate that trees for coppicing need to be established only once and can then be used for many years. Farmers are obliged to manage trees well at all times during the cropping season so that tree and crop competition is minimised. Another constraint for wider use of the technology in the region is the limited availability of tree seed, especially for *Gliricidia*, but farmers will gradually overcome this constraint as they increasingly propagate this from stem cuttings. Livestock do not browse fresh *Gliricidia* leaves, and therefore trees need little protection during the dry season. (For a more detailed description of mixed cropping of *Gliricidia* and maize see Ikerra *et al.* 1999.)

Annual relay cropping of trees

Nitrogen-fixing trees are planted into a field at a time when crops have already been well established. Trees such as *Sesbania sesban* and *Sesbania macrantha* are first raised in nurseries, then bare-rooted seedlings are transplanted into the field. Species such as *Tephrosia vogelii* or *Crotolaria* spp. are sown directly under a canopy of established crops. The trees thrive mostly on residual moisture and develop their full canopy only after the crop is harvested. As farmers prepare land for the next season, they clear-cut trees and incorporate the biomass into the soil, and then repeat the cycle. Trees must thus be replanted every year.

As with coppicing, the main objective is to replenish soil fertility and achieve household food security in land-scarce farming systems. The main limitation of the technology is that farmers must depend on late rainfall for trees to become well established. In very dry years, there is a high risk that trees will produce little biomass and hence have little effect on crop yield. Labour needed every year for establishing the trees could be another constraint, although less labour is needed for species that are sown directly. Yield increases are less dramatic than with the former two technologies, because the trees produce less biomass.

Trees that are browsed, such as *Sesbania*, need to be protected during the dry season in areas where livestock grazing is not regulated. (For a more detailed description of relay cropping with *Tephrosia* see Böhringer *et al.* 1999b.)

Fodder banks

A fodder bank is a small, well-protected, and intensively managed plot of trees that is cut continuously for feeding livestock. Species with high nutritional value are preferred such as *Leucaena* spp., *Calliandra calothyrsus*, and *Acacia angustissima*. Tree spacing may be as close as 1 x 0.5m, but it may be wider where fodder banks are intensified by planting fodder grasses such as napier between tree rows. Fodder banks are usually planted close to the place where livestock are kept in order to minimise the amount of labour for carrying the fodder. Many smallholder dairy farmers are currently using this technology, but it also has potential for milk goats and, possibly, other livestock (see also Wambugu *et al.* 2001 and in this collection).

The main objective of fodder banks is to increase the income of smallholder dairy farmers by substituting the fodder for expensive, purchased feed concentrates and by increasing overall milk yield, especially during the dry season, when fodder from natural grazing sources becomes extremely scarce. Access to markets for milk is a precondition for the technology to be profitable, and hence farms in a peri-urban setting have a comparative advantage. Trees can be cut just one year after planting, reach their full potential in the second year, and then be continuously harvested for many years. Processing and storing the tree fodder on-farm offers considerable opportunity for adding value.

The main limitations of the technology are the labour needed to establish the trees well in the first year, and the need for solid fencing throughout to protect them from roaming livestock. Major constraints in the region are that improved dairy animals are scarce and generally unavailable, and that small-scale farmers who want to buy animals lack access to loans.

Rotational woodlots

Rotational woodlots are normally small plots of trees (0.04–0.5 ha), which are well-protected, particularly during the first two years after they are planted. Tree species planted by farmers are Australian acacias (*A. crassicarpa*, *A. julifera*, *A. leptocarpa*), native acacias such as *A. polyacantha*, neem (*Azadirachta indica*), and *Senna siamea*. Trees are usually planted 2 x 2 m apart, and farmers often use the spaces in-between for intercropping with a crop such as maize during the first two years that the trees are growing. In the third and fourth years, when trees have reached a height beyond the reach of livestock, and intercropping has to cease because of the shade created by them, animals are allowed to enter for grazing. Trees are clear-cut in year four or later, after which soil fertility is also restored where nitrogen-fixing acacias were planted.

The main objectives with rotational woodlots are to make households self-sufficient in wood for fuel and construction, and to provide some additional income through the sale of wood. Rotations of four to five years are possible where land is less limited and farmers can allow at least two years of fallowing. Where land is scarce, farmers have adapted the technology to boundary plantings. The technology has the potential to produce 60–80 tonnes/ha of dry wood five years after planting compared with an average productivity of natural *miombo* of approximately 32 tonnes/ha (Ramadhani *et al.* in press). It may thus offer an economic alternative to ongoing deforestation of the *miombo*, particularly in areas where fuelwood is in high demand for activities such as curing tobacco.

The main limitations of rotational woodlots are the long period (four to five years) farmers have to wait for the technology to provide wood products, the high labour costs during the first year of establishment (including for protection), and the lack of water for nurseries in drier areas such as in Tanzania, where it has particularly large potential in the heavily deforested Shinyanga and Tabora regions. Furthermore, an extreme scarcity of tree seed for Australian acacias inhibits wider spread of the technology. (For a more detailed description on rotational woodlots see Ramadhani *et al.* in press.)

Planting indigenous fruit trees

Individual fruit trees are planted as boundaries along field edges, on contours, or around homesteads. They are usually well protected and looked after, with some occasional watering needed during the first dry season after transplanting. Farmers' priority species from the *miombo* are *Uapaca kirkiana*, *Sclerocarya birrea*, *Strychnos cuccloides*, *Parinari curatellifolia*, *Vangueria infausta*, and *Azanza garckeana*, all highly valued for their nutritious fruit.

The main objective of planting indigenous fruit trees is to safeguard the nutritional security of the family, in particular children, since all indigenous fruits are high in sugars, vitamins, and minerals, and many trees are in fruit during the seasons when people often go hungry.

They also provide farming families with income through the sale of fresh fruit, a potential that could be further developed by promoting processing and marketing. Another objective is to conserve biodiversity of the dwindling *miombo* tree resources. The true merits of planting indigenous fruit trees still need to be determined through researching the markets and product development.

The main limitation now is lack of knowledge on the best propagation techniques. Research on the best methods for on-farm planting is still in its infancy, but some success in germinating seed and in vegetatively propagating the plants has been made, especially with marcotting (a propagation technique involving inducing roots to grow on a small branch while it is still attached to the larger tree). If the time to first fruiting of these species could be reduced to well below five years, the economics of planting on a larger scale would certainly be improved.

Linking agroforestry innovations to development

Since 1992, on-farm research has become the main vehicle for assessing the biophysical and economic performance of these technologies, with farmers gradually taking over the design and management of trials in their fields. By the 1996–7 season, approximately 5000 farmers participated in this kind of research across the four countries, most of the testing being on improved fallows in Zambia (Kwesiga *et al.* 1999). On-farm experiments are usually characterised by intensive farmer–ICRAF interaction, and individual farmers are supported with information, training, and technical visits. Researchers provide a lot of this support during field visits for data collection.

Here, the support given to individual farmers could be considered as the minimum incentive necessary for making agroforestry technologies adoptable in the first place. Such incentives are expected with agroforestry, which is classified as a preventive innovation (Rogers 1995), meaning that the time from tree planting until tangible benefits accrue is relatively long. For instance, significantly increased maize yields after a two-year fallowing with *Sesbania* in on-farm trials created a lot of enthusiasm among peer farmers, which again triggered a large demand for the technology in Eastern Province of Zambia (Kwesiga *et al.* 1999). This highlights the fact that disseminating these agroforestry technologies has now evolved into a more client-driven process. But this change occurred only after a good number of first-time testers demonstrated the benefits in their fields to their peers, who could see the results for themselves.

The number of farmers across the four countries who are using the new agroforestry technologies outside on-farm research arrangements with ICRAF indicates the demand – they totalled approximately 30,000 in the 1999–2000 season, of whom more than 40 per cent were women.

Approaches to accelerating impact: agroforestry as a learning tool

Scaling out through partners

To achieve impact, our main strategy focuses on working through existing government, non-government, and development organisations and farmer groups. This scaling out aims to influence partner organisations and their policies through networking, lobbying, and collaboration (Scarborough *et al.* 1997) so that we can achieve our goal of catalysing a client-driven wider use of agroforestry technologies in order to improve rural livelihoods in the region significantly. Partner and ICRAF contributions in Southern Africa in this scaling out vary considerably (see Table 3), but this diversity is needed to involve mainstream development agents. The institutional and managerial set-up of government and non-government agents is distinct. The former is commonly more hierarchical with top-down planning and implementation, while the latter tends to have better grassroots participation yet is often weak in integrating development efforts systematically into existing structures and hence in providing impact beyond project areas.

In collaborating with such contrasting partners, we want to compare successes and failures and assess transaction costs in partnership, which should lead to a better understanding of which are the most effective ways of scaling out agroforestry. At the same time, we hope that hybrid diffusion systems may emerge (Rogers 1995) that will successfully combine existing technology transfer by national extension services with participatory, decentralised innovation processes happening locally. This implies that parts of our collaboration must interact with farmer groups, which gives us an opportunity for direct client consultation. Thus the quality of our core services (provision of information, knowledge, tree seed, and capacity building) can continuously improve through feedback from farmers. ICRAF's overall role in Southern Africa is therefore one of a facilitator between government research-extension services, which continue to operate

Table 3: ICRAF's operational modes with partners in Southern Africa		
Partner	**Partner contribution**	**ICRAF contribution***
Government	• Infrastructure • Executive power • Personnel • Tax rebates	• Germplasm • Scientific knowledge • Networking • Capacity building • *Operational funds*
NGOs	• Grassroots-level organisation • Personnel • Operational funds • Practical feedback	• Germplasm • Science/knowledge • Networking • Capacity building • *Institution building* • *Empowerment*
Farmer groups	• Land • Time • Labour • Indigenous knowledge	• Germplasm • Science/knowledge • Networking • Capacity building • *Compensation* • *Organisational support* • *Empowerment*

* ICRAF contributions in italics are those that need to be provided in addition to the core services (germplasm, science and knowledge, networking, and capacity building) to make collaboration more effective.

largely on the linear model of technology transfer and local, decentralised extension processes of participatory technology development. Furthermore, a dialogue between participatory technology development actors and formal extension and research institutions is also facilitated, providing opportunities for feeding research hypotheses from the grassroots level into the formal research set-up.

At present, we are collaborating with 572 partners in the four core countries, comprising government agencies (36), development projects (16), NGOs and grassroots organisations (26), farmer groups (485), and the private sector (9). We have extended our activities into Mozambique through the Danish International Development Agency (DANIDA), the German Gesellschaft für Technische Zusammenarbeit (GTZ), and World Vision International. In South Africa, we have conducted joint training activities in former homelands with the Danish Cooperation for Environment and Development (DANCED) in Mpumalanga and with the Finnish International Development Agency (FINNIDA) in Northern Province.

The main instrument for collaboration in these four countries is open, informal, biannual 'networkshops', which ICRAF facilitates. In these workshops, representatives from partner organisations and farmers together plan and review the implementation of agroforestry activities. A series of 'networkshops' were used to develop detailed operational plans until March 2001 and strategic options beyond (Böhringer *et al.* 1999a). Six main operational objectives were identified as necessary for overcoming common drawbacks:

- increasing the awareness of stakeholders, including farmers;
- strengthening the capacity of farmers and extension agents;
- amplifying the availability of germplasm at the grassroots;
- improving partnership and co-operation among stakeholders;
- supporting the marketing of tree products;
- institutionalising participatory approaches and methods for innovation development and extension.

ICRAF has given the first four points much attention since it first engaged in development in Southern Africa in 1997. They are action oriented and seek to prepare a platform that will help broaden the impact of agroforestry in the region. The first one is to overcome the limited awareness that stakeholders have, including farmers, of agroforestry potential and benefits. Awareness is increased by using common dissemination tools such as holding field days (reaching on average 2500 farmers, about half of whom are women), in each of the four countries every year, and producing and distributing agroforestry extension materials (leaflets, manuals, cartoons, posters, radio programmes, and videos) and regional and national newsletters.

Grassroots capacity can be strengthened by helping farmers to form groups, facilitating direct farmer-to-farmer training in villages, training farmer trainers who will lead community-based extension, and providing technology-related skills training on topics such as how to manage nurseries and trees. We have found that this 'farmer first' approach to capacity building is efficient. Farmer-to-farmer training, for instance, costs on average approximately US$2.50 per farmer trained, lasting usually three to four days in villages. In comparison, a one-week residential training course costs on average between US$20 and US$30 per farmer. One trained farmer typically reaches up to ten other farmers during the first season after training in agroforestry.

Supporting decentralised grassroots-level germplasm production and building supply networks help to address the problem of restricted availability of germplasm, mainly tree seed. Here, the projections until the end of year 2001 are that ICRAF will help to establish 800 farmer seed-multiplication plots and over 6000 farmer nurseries (Böhringer *et al.* 1999a). Partnership and co-operation among stakeholders can be improved through the scaling-out process described earlier.

Our facilitation role in marketing tree products is a more recent one, as we appreciate more and more the need to improve links between small-scale farmers and markets. Thus, the development of innovations and the use of new technologies are ultimately driven by consumer demand. This is particularly true for technologies for which generating income is an important objective, such as cultivating indigenous fruit trees or fodder banks. In promoting indigenous fruit in particular, the need to assess market demand and consumer preferences is immediate; therefore, links need to be established between producers and markets. Building these links must start before large numbers of trees planted start bearing fruit. Here, experience from the tree-crop sector is particularly valuable, and ICRAF in Southern Africa collaborates closely with a GTZ-funded regional project, 'Integration of Tree Crops into Farming Systems'. This project has put into place in Kenya a successful model for exotic fruits such as mango, papaya, and cashew. It integrates product development, processing, capacity building of farmers, and farmer-to-farmer extension in a holistic way (Van Eckert 1997). The challenge for the years to come is to adapt this approach for indigenous fruit to Southern Africa and to draw more private partnerships into the network, particularly from South Africa, where markets are well developed.

Pilot development projects: understanding impact under real-world conditions

Keeping our goal in mind we need to ask: how much of the technology developed by research actually reaches the farming world through the technology-transfer approach? This approach still predominates in extension services in the region. Technology transfer produced remarkable impacts during the Green Revolution in parts of Asia, largely by limiting its focus to interventions that targeted homogeneous cropping systems with large geographic spread such as irrigated rice; but it has failed to show similar impact for the large majority of smallholder farming systems elsewhere.

In Southern Africa, biophysical conditions are more marginal and socio-cultural settings more variable, which calls for the development of diverse and often complex agricultural systems that meet farmers' multiple needs. These complex and diverse technology innovations are best developed locally. Only continuous farmer experimentation and adaptation *in situ* will make them feasible and profitable to farming communities at large – the ultimate clients of research and development services.

This calls for a better understanding of the complex inter-dependencies among the major biophysical, economic, social, and cultural factors that come into play at a larger community or even watershed scale, so that the process of agroforestry innovation, development, and extension can be conceptualised for wider scaling up. Communities drive this process, and technical options are primarily perceived as learning tools that bring about social change. Only to a lesser extent are they seen as linear vehicles for increasing adoption rates of technical 'solutions' *per se*.

ICRAF therefore considers that establishing pilot development projects is an important link between research and development. In them, hypotheses in natural resource management and agroforestry development can be tested together with communities. Results can be assessed in a participatory way and documented more holistically at a landscape level. We are currently selecting and designing at least one pilot project with partners and communities in every core country. In Malawi, three communities have been involved in a watershed near Zomba since 1998. Our working hypothesis there is that agroforestry makes conservation farming on the steep slopes more profitable and catalyses the effective conservation of soil and watershed in the community.

The approach we follow is to facilitate community dialogue based on the principles of adult education and 'critical consciousness'; that is, the process of reflection and action needed for a community to improve itself (Freire 1969). The approach starts with training for transformation; community action follows, with conserving fields and establishing farmer nurseries; village monitoring, and planning workshops then follow on. Such cycles of community reflection and action are repeated and are gradually expected to increase community capacity to higher levels so that development becomes sustainable.

Already, after two years, work in this pilot development project has given us some crucial insights into researcher–farmer communications, gender–wealth relationships, and community organisation. Outside experts tend to oversimplify scientific information, while farmers are asking for comprehensive and scientific explanations for natural processes to be given in their own language and in terminology based on indigenous knowledge.

Annual household incomes in the area range from as little as US$80 to US$500, yet villagers clearly distinguish four wealth tiers. Peer communication happens mainly within these four wealth tiers and a gender aspect in communications cross-cuts them, meaning, for instance, that women within the same wealth group communicate most readily with each other, and then with women in different wealth groups, but rarely with men, and vice versa. Organisation of groups follows a similar pattern.

The challenge will be to draw in more women, who head 46 per cent of all households in the area, and more of the poorest, but who have little means to participate in ICRAF-facilitated activities as they are too busy with their own household problems and cannot risk trying out innovations. Here, community soil and watershed conservation is thought to provide an excellent tool for facilitating social change, as its effectiveness relies on being all-inclusive, which means that the wealthier members in the community have to find ways to assist the poorer ones with conserving their fields so that everyone in the watershed can benefit. Our strategy for scaling up such pilot development projects is to help make direct farmer-to-farmer links within and across countries in the region. To do this, we work with projects such as those of Oxfam GB in the Mulanje area in Malawi, which has similar objectives and uses approaches like those in our pilot development project.

Monitoring and evaluation: a key element in the learning process

The focus of our work over the past three seasons has been on getting agroforestry action initiated with multiple partners. Activities have largely been driven by demand and supported by a very limited number of ICRAF staff in Southern Africa. Therefore, our monitoring and evaluation (M&E) efforts so far have had to concentrate on measuring outputs quantitatively using conventional methods such as questionnaires and record sheets, for example, to characterise households or to capture information about farmer nurseries. These efforts have been project driven, and little dialogue has taken place so far on what M&E means to different stakeholders, including farmers, how it can be approached together, and how it can be used as a central learning tool within our network. However, as people add value to information obtained from monitoring, we must find answers to important questions about who is evaluating, why, and for whom (Guijt *et al.* 1998).

This kind of dialogue should lead to more participatory methods being used in M&E. It would also necessitate finding a common language for communicating among farmers, scientists, project administrators, fieldworkers, donors, and others, so that agreement can be reached on who wants to evaluate what, with whom, and how. We envisage, therefore, a practical approach to M&E, which could be founded on three pillars: farmer self-M&E, external (conventional) M&E by development agents, and village impact assessment workshops in selected representative communities. With all three pillars, stakeholders would participate in identifying impact indicators and the design, implementation, and analysis of common M&E tools. Triangulation among results from these three approaches should be a more transparent, accurate, and cost-effective way to reflect successes and failures of our work and to advance our learning (Guijt *et al.* 1998).

Still, the main way in which participation can add value should be that it builds grassroots capacity in situation analysis and that it empowers people to reflect on actions in a structured way before new activities are planned and implemented (Freire 1969). ICRAF wants to catalyse this process, with the main aim being to use participatory M&E as a tool for planning and learning within our network. This would also elevate M&E from an internal project operational output to a developmental output with larger benefits to the public. Dialogue has been initiated since the 1999–2000 season with a series of village workshops, followed by a 'networkshop' focusing on M&E. The objective was to understand farmers' own concepts of M&E and priorities better, and to identify key indicators for joint assessment. This process has just begun, and first experiences will be evaluated in another 'networkshop'.

Some lessons and preliminary conclusions

Getting the right partnerships

After three years of engagement in development in Southern Africa, ICRAF has reached out to a considerably larger number of farmers compared with earlier forms of on-farm research. This has been achieved mainly through networking and through scaling out agroforestry to an increasing number of partners and collaborators in the region. Partnerships have now become so numerous that the costs of handling them stretch ICRAF's limited resources and capacities to the maximum. Therefore, it is now time to analyse partnerships and identify the efficient ones that contribute substantially to our reaching our goal. For example, a number of partners such as NGOs and bilateral development projects are willing to dedicate some of their resources to agroforestry, while others are not, yet they all want our continued support.

Special cases are government partners in all countries, who have scarcely any operational funds at their disposal. The large amounts of funds that government partners put forward for field activities are usually allocated to paying staff allowances, which are meant to supplement very low government salaries. Such staff allowances may typically total more than 60 per cent of overall costs for a planned joint activity. This raises questions about the appropriateness of scaling out under such conditions. While we appreciate the importance of government extension services in providing a national institutional umbrella, their role needs to be redefined. They should move from their current poor delivery of services to a role of facilitating and coordinating services – and this ideally with minimum overhead costs. Yet, currently, government facilitation for local development agents and grassroots movements remains marginal in most cases, while vertical integration of these development efforts into national strategies and institutions hardly exists.

This lack also explains to some extent why a fragmented multitude of multilateral and bilateral development projects, NGOs, and charities, most of them financed externally, are largely driving local development in Southern Africa today. This situation cannot be sustainable in the long run. ICRAF's role as a facilitator is therefore seen to be a limited and temporary one, since national institutions should naturally take over this role. Whether this changeover succeeds will also depend on policies that favour the decentralised development of innovation and demand-driven delivery of services to smallholder farmers. The latter will probably need more public and private partnerships in the region to become effective.

The timeframe for agroforestry impact

Another experience with scaling out is that, aside from pilot development projects, it allows ICRAF only limited direct assessment of impact in the communities. Partners' capacities in M&E are overall weak, and the need for M&E is often dictated by external donors, who very seldom put natural resource management or agroforestry high on their agenda. An assessment of the overall effect of scaling out on the actual impact of agroforestry thus becomes very difficult, if not impossible. Still, we are reaching as many as 10 per cent of households with agroforestry in a given area in a relatively short time. But this achievement also depends on the combined effort and resources that partners and ICRAF spend.

Such a reach conforms well to results from other diffusion studies reporting that 'innovators' typically account for 2–3 per cent of the total number of members of a given social system and 'early adopters' 12.5 per cent (Rogers 1995). However, agroforestry needs special consideration, because innovation–decision periods of two or three years are common. This is the period that elapses between acquiring awareness and knowledge about the potential of a technology and its actual use or adoption. The earliest it will happen in agroforestry is when tangible benefits accrue from trees for the first time. It is therefore far too early to predict the likely outcome of continued scaling out or, if more project resources were available to ICRAF, of scaling up agroforestry in the region (Scarborough *et al.* 1997). Either strategy, or both combined, should eventually lead to a 'take-off stage' in technology use, where as many as 30 per cent of the households are reached (Rogers 1995).

For now, we consider bridging the decision–innovation time period of two to three years to be crucial, which means that our main efforts are focused on steering the current first generation of agroforestry users towards success. The numbers should greatly increase in Malawi, Tanzania, Zambia, and Zimbabwe after the 2000–01 season.

Farmers as key agents of change

Our collaboration with large numbers of farmer groups and communities often entails involving limited numbers of farmer representatives for specific activities, after which they should act as agents of change in their home areas. Here, we experience again and again how important and yet difficult is the proper selection of farmers. Our partners select the farmers to represent their communities in joint activities such as field days or farmer-to-farmer training. Later, when we interact with the communities concerned more closely, we often learn that the roles, responsibilities, and privileges of persons who were selected to represent groups or whole communities were not discussed openly, and that little agreement was reached on the selection criteria to be used. Indications are that the personal criteria of extension agents also often govern farmer selection.

Farmer 'representatives' selected on such a basis are likely to act only with great difficulty as agents of change in their home communities. They experience 'social levelling', meaning they lose influence and become ordinary farmers again within a short time instead of evolving into leaders. It is, therefore, of utmost importance that farmer representatives be selected with a true mandate from their communities, and endorsed by traditional leaders. This process may take lengthy facilitation and negotiation, but we learn that this is time well spent in order to avoid unwelcome social disruptions during a period of change that is difficult anyway as innovations are introduced into the community.

Addressing the special needs of women

Getting the right gender balance in our work has proved more difficult than originally anticipated. Part of this problem is inherent in partner institutions, with, for instance, fewer than 3 per cent of extension staff working directly with farmers in the region being women. Without involving more qualified women, especially where impact is to take place, it appears unlikely that gender barriers in communications can be easily overcome, or much progress made in increasing the use of agroforestry among the large numbers of female-headed households in the region. One way in which ICRAF tries to overcome this discrepancy is by ensuring that at least half of those who participate in any joint activity must be women. This is one of the few conditions that we place on collaborating partners. With this condition, we hope gradually to build grassroots women's capacity, so that this marginalised group can eventually gain more influence in ongoing decentralisation processes in the region.

Agroforestry – first and foremost a learning tool

In summarising our work to date, the most important lesson is that agroforestry, because of its inherent complexity and diversity, emerges as a central learning tool in building local capacity to develop innovations. Secondly, agroforestry offers farmers technology options that can significantly contribute to improved livelihoods. We particularly see a great need for many dispersed local innovation centres to emerge, which should then be linked horizontally to achieve wider impact. This calls for some new thinking within ICRAF and other international agricultural research centres that have promoted a research paradigm focused on developing models with the widest possible applications based on simplifying the complex, separating the connected, and standardising the diverse. It appears unlikely that models for agroforestry development can be constructed in a similar way, as the variability of the biophysical and socio-cultural settings of smallholder farmers is very great in the real world. However, pilot development projects have been shown to be crucial, because they provide ICRAF in Southern Africa with opportunities to learn, even as we develop innovations, about what can happen at a wider landscape or watershed level. This enhances our own understanding of critical factors that can lead to success – or failure – in our efforts to facilitate development in the region through agroforestry.

Acknowledgements

The author is grateful for the financial support given to ICRAF and for carrying out the work described in this paper by the Canadian International Development Agency (CIDA) through funding the project 'Agroforestry for Sustainable Rural Development in the Zambezi River Basin, Southern Africa Region' (R/C Project 050/19425, Agreement 23591). I would also like to thank Peter Cooper and Steve Franzel for reviewing an earlier version of this paper.

References

Böhringer, Andreas (1999) *Dissemination and Development for Accelerated Impact: Our Strategies in Southern Africa*, Makoka, Malawi: ICRAF

Böhringer, Andreas, Roza Katanga, Prince R. Makaya, Nobel Moyo, and Stephen Ruvuga (1999a) *Planning for Collaboration in Agroforestry Dissemination in Southern Africa*, Southern Africa Agroforestry Development Series No. 1, Makoka, Malawi: ICRAF

Böhringer, Andreas, Jumanne A. Maghembe, and Rebbie Phiri (1999b) '*Tephrosia vogelii* for soil fertility replenishment in maize-based cropping systems of southern Malawi', *Forest, Farm and Community Tree Research Reports* 4: 117–20

Freire, Paulo (1969) *Education for Critical Consciousness*, New York: Continuum

Guijt, Irene, Mae Arevalo, and Kiko Saladores (1998) 'Tracking change together', *PLA Notes* 31: 28–36

Ikerra, Susan T., Jumanne A. Maghembe, Paul C. Smithson, and Roland J. Buresh (1999) 'Soil nitrogen dynamics and relationships with maize yields in a gliricidia–maize intercrop in Malawi', *Plant and Soil* 211: 155–64

Kwesiga, Freddie and Jan Beniest (1998) *Sesbania Improved Fallows for Eastern Zambia: An Extension Guideline*, Nairobi: ICRAF

Kwesiga, Freddie, Steve Franzel, Frank Place, Donald Phiri, and Percy Simwanza (1999) '*Sesbania sesban* improved fallows in eastern Zambia: their inception, development and farmer enthusiasm', *Agroforestry Systems* 47: 49–66

Ramadhani, Tunu, Robert Otsyina, and Steven Franzel (in press) 'Improving household incomes and reducing deforestation: the example of using rotational woodlots in Tabora District, Tanzania', *Agriculture, Ecosystems, and the Environment*

Rogers, Everett M. (1995) *Diffusion of Innovations*, 4th edition, New York: The Free Press

Scarborough, Vanessa, Scott Killough, Debra A. Johnson, and John Farrington (1997) *Farmer-led Extension: Concepts and Practices*, London: ITDG Publications

UNDP (1999) *Human Development Report 1999*, New York: OUP

Van Eckert, Manfred (1997) 'Integration of tree crops into farming systems project in brief', paper presented at the ICRAF–BMZ Workshop on Trees and People, Bonn: Bundesministerium für wirtschaftliche Zusammenarbeit und Entwicklung

Wambugu, C., S. Franzel, P. Tuwei, and G. Karanja (2001) 'Scaling up the use of fodder shrubs in central Kenya', *Development in Practice* 11(3–4): 487-94

Scaling up participatory agroforestry extension in Kenya:

from pilot projects to extension policy

T.M. Anyonge, Christine Holding, K.K. Kareko, and J.W. Kimani

Current trends in extension and expected components of extension approaches

In the past, public-sector agricultural extension and research services in developing countries played a vital role in promoting technological innovation in agriculture. However, changes in the structure of the public sector, the context in which it operates, and the likely nature of future technological innovation raise questions about whether these institutions will be able to meet the challenge of the continued need to increase agricultural productivity. Over the last decade or so, therefore, several attempts have been made to establish agricultural services that are responsive to resource-poor farmers. In most of these experimental programmes, farmers, rather than professional extensionists or researchers, have acted as the principal agents of change (Scarborough 1996; Farrington 1998). In the current context of market liberalisation and deregulation, small farmers are initiating and implementing significant adaptation strategies, which include diversifying to new market niches, contracting agriculture with agro-industries, and forming local collective organisations for marketing and post-harvest activities, as well as engaging in more off-farm employment (Berdeque 1998; Ellis 1999).

This paper shares the experiences of implementing a natural resource extension programme, highlighting three innovative components of an extension project on managing natural resources. The programme was run in Kenya from 1990 to 1998 by the Kenyan government and the Finnish International Development Agency (FINNIDA). It comprised three components, namely:

- assessment of the impact of conventional service delivery;
- development of participatory extension methods, such as local planning;

- incorporation of the experiences of pilot participatory extension projects into national extension policy (scaling up).

Both governments and NGOs provide research and extension services. The increased interaction between farmers and these providers is described as follows:

- *pluralistic* – incorporating service providers from the private sector, churches, NGOs, community-based organisations (CBOs), and conventional service providers;
- *integrated* – addressing production issues on the farm in an integrated cross-sectoral manner that responds more closely to farmers' own perceptions of on-farm interactions and decision-making;
- *bottom-up* – participatory, farmer-led, gender-aware, and empowering; in other words, farmers plan, design, and lead the extension process, and efforts are made in extension planning to be as representative as possible of the various social institutions in a community.

Congruently, extension services throughout sub-Saharan Africa (Malawi, Uganda, Zimbabwe, to name but three countries) are going through a period of radical transformation, actively seeking to institutionalise participatory planning processes. The door is open for contributions of practical innovative approaches (for example, Veldhuizen *et al.* 1997) that can be sustained within and spread between communities. However, to facilitate more responsive planning of extension services, we need greater understanding of local processes of institutional, political, and economic change, with which to inform a more judicious selection and application of participatory methods (Mosse 1994).

The Nakuru and Nyandarua Intensified Forestry Extension Project

The Nakuru and Nyandarua Intensified Forestry Extension Project, or *Miti Mingi Mashambani* (Swahili for 'many trees on farms'), began in October 1990. It was jointly funded by the Kenyan government and FINNIDA[1] and implemented by the Forestry Extension Services Division. The development objective of the project was to sustain the supply of essential tree products and to stabilise and improve the rural environment through general afforestation.

The project was divided into three components: training; logistical support; and improved extension:

- *Training* – this component provided training in agroforestry, extension, and communication skills for forestry extension staff at all levels, from national to village. The project also provided training for staff from collaborating ministries at district and divisional level, chiefs, and subchiefs. This component had the substantive task of developing a team spirit and raising morale among extension officers who regarded the transfer from plantation to extension forestry as a demotion.

- *Logistical support* – this component constructed and established offices, supported a few institutional nurseries, provided transport (motorbikes and bicycles), and supplied germplasm for establishing on-farm nurseries.

- *Improved extension* – this component concentrated on improving and intensifying the existing conventional extension approaches that were being implemented with schools, groups, and contact farmers.

With the school approach, for example, components included roof-water harvesting, establishment of tree nurseries for training, teacher training, parent–teacher association seed stands for the surrounding community, school environment clubs, school open days, woodlot establishment, and installation of improved institutional stoves for better use of fuelwood in boarding schools. Schools are inappropriate venues for the mass production of seedlings because supervision and watering are intermittent, with school holidays falling at crucial times in seedling production. However, school nurseries and agricultural compounds proved excellent venues for community-focused training and method demonstrations (Niemi 1995).

The improved extension component of the project was also mandated to pilot new extension methods and approaches that would improve the effectiveness, impact, and relevance of extension. At the beginning of the project in the early 1990s, the 'Training and Visit' system, though largely discredited (Antholt 1994; World Bank 1994; Carney 1998), continued to predominate as the extension approach in Kenya. The new methods being piloted sought to develop approaches of integration and participation. These represented the earliest attempts the Forestry Extension Division made to address these issues and seek operational ways to include them in their extension programmes. The issues raised in implementing these new approaches contributed significantly to the discussions of the role of forestry extension services in the broader context of providing agricultural extension services in the country. The two methods piloted were farmer-designed trials (Franzel *et al.* 1996) and local planning.

Enhanced implementation of conventional service delivery

Two methods were used towards the end of the project to measure the effectiveness of earlier agroforestry extension activities.

An assessment of the effectiveness of the conventional extension channels – contact farmers, schools, and groups – was conducted in 1995; it covered 216 farming households selected in a two-stage sampling process. The contact points were classified into three agro-ecological zones, and households were selected in four directions at a distance of up to 2 km from the contact points. The participatory component of the survey used focus-group discussion with partici-pating and non-participating farmers to better understand the dynamics at contact points. The results showed that schools were the most effective mechanism for outreach in Nakuru (reflecting the heavy investment in this channel as an extension medium in Nakuru District), while in Nyandarua, groups were the most effective channel. In both districts, contact farmers were the weakest and least effective channel.

A second method used for assessing impact was on-farm surveys of woody biomass. Surveys based on aerial photographs were conducted in 1993 and 1998 to assess changes in farm woody biomass resulting from project interventions. In an intensive aerial survey made in 1993, the sampling unit of the inventory was the farm, and some of the sampled farms were visited in 1993 and again in 1998. The data collected covered planting niches, tree species, origin of germplasm, trunk diameter, and projected end-use. Between 1993 and 1998, the useable volume of wood per farm in the project area rose from 7.5 to 17.07 m³. This latter exceeded by 12 per cent the calculated annual requirement per household of 15 m³, made from the project's socio-economic survey in 1991 (Holding and Kareko 1997; Hoyhta *et al.* 1998; Njuguna *et al.* 2000). However, reliable interpretation of results from such a survey can be made only if contextual information is available such as settlement patterns, land tenure, germplasm availability, and tree use. In the two phases of the project, several socio-economic and marketing studies were conducted to obtain this contextual information. These provided in-depth analysis of farmers' decision making and complemented the findings of the survey. In this case, data from the woody biomass survey analysed in conjunction with other project data demonstrated that the development objective of the project had been achieved.

What were the elements of success?

Several interrelated factors contributed to the achievement of the project's objectives:

- The project area was a settlement area, largely devoid of trees. Farmers settling in the area were keen to indicate the boundaries of their farms, establish privacy, and protect their houses from the strong winds.

- During the 1980s, in an effort to curb the felling of indigenous trees as agricultural areas expanded, a tree-felling permit was introduced. This permit was interpreted by the administration to apply to all trees on farmland. The process in time and money required to obtain this permit often exceeded the value of the trees to be felled. This was a considerable disincentive to tree planting. The project actively sought to have the provincial administration, which was the enforcement agent, declare the tree-felling permit redundant in the project districts, and it succeeded in doing so.

- The project facilitated the supply of germplasm. Initially it went directly to farmers; later it was supplied through community and farmer seed stands. The project also incorporated training in seed production, distribution, and handling.

- Training for all stakeholders – farmers, extension staff, administration, policy makers – was regular and frequent.

- Extension access was reinforced, as the programme worked through contact farmers, groups, and schools. In this way it reached and interacted with each member of a household: men, women, and children. This reinforcement of access had considerable impact on the willingness of households to experiment with agroforestry.

- Training and message reinforcement led to a change of attitude among staff and farmers. Staff had previously been sceptical about the roles of extension and agroforestry. Farmers had previously assumed that the 'government will provide'. Crucial to the success of the programme was the raising of morale in the extension service and the fact that farmers were empowered to test and develop agroforestry interventions.

- Extension approaches and agroforestry technologies were selected to match the specific site requirements and socio-economic context of the communities. For example, in Nakuru, as the farms were small and located near markets, high-value trees were in demand and the project supplied them at cost. As the farms were small, trees with fuelwood as a by-product had to be compatible with crops – hence there was a high demand for *Grevillea robusta*. In Nyandarua, where the farms were larger and the climate colder, there was a demand for eucalyptus for woodlots and windbreaks. For timber production, *Cupressus lusitanica* was popular. In every district, soils, altitude, and climate affected choice of species and technology. Between farms there was also variation pertaining to external remittances and life-cycle trends, affecting both potential for investment and cash needs. Thus blanket recommendations were not encouraged; the extension service instead offered a range of tree species and technology components from which farmers could select and adapt to suit their particular needs and situations.

Piloting participatory extension approaches

Experience with local-level planning

Local-level planning (LLP) used participatory rural appraisal (PRA), a tool regularly used in the extension services and NGOs in Kenya, but LLP went further in implementing, monitoring, and evaluating with the community. Pilot activities conducted under the auspices of the Nakuru and Nyandarua Intensified Forestry Extension Project tried out extension methods in which farmers remained the central figures during planning, implementation, and monitoring of their development activities.

Local-level planning was conducted as a pilot project in two administrative locations of Nyandarua District – Subego in Ndaragwa Division and Weru in Ol Joro Orok. During this time, the Ministry of Agriculture's 'catchment approach' in soil and water conservation was being implemented, which used participatory rural appraisal (PRA). Participatory approaches were also being piloted by NGOs and in donor-financed projects. These participatory and integrated extension services were being piloted as an alternative to the failing 'Training and Visit' system.

The LLP activities started with forming multi-agency divisional extension coordination committees in the two locations, and training them in participatory methodologies. This training was followed by open village meetings (or *barazas*) during which presentations of the forthcoming process were made, and farmers were requested to select villages for focused group discussions in agroforestry assessment surveys, which would be based on PRA principles. These surveys were guided by a checklist focusing on the natural resources sector (Tengnäs 1992) in order to ensure a focus on issues within the project's mandate.

A community action plan was developed, and the various ministries committed themselves to activities to be implemented through a memorandum of understanding. Monitoring and evaluation of the activities with the community took place in an open *baraza* in Subego and in a meeting with village elders (community representatives) in Weru. In the meeting between technical officers and elders, it was possible to set priorities on issues jointly and allocate available funds to activities in a transparent manner.

Activities requiring inputs from outside the location were financed on a cost-sharing basis with the farmers. In constructing water jars, for example, the project worked through women's groups. Each group wished to build a water jar for each member's compound. This was initiated by monthly contributions from each member to build up the group capital. The project provided the materials not available locally – cement and chicken wire – and the group provided sand, ballast, and labour. Thus the cost of each jar (US$30) was shared roughly equally. In the first year of activity, despite a jointly planned estimate of five water jars, 22 were constructed.

Community training relating to identified activities formed a major component of the community action plans. Activity monitoring was conducted through quarterly meetings – again, in *barazas* in Subego and meetings with village elders in Weru.

Key observations

The observations cited should be viewed in the context of the dominant extension paradigm at that time. Participatory extension approaches had not been institutionalised in the early 1990s. Listed below are findings learned in implementing this pilot activity:

- When farmers lead in the extension process, implementing and monitoring any jointly developed work plan is more straightforward and resources are distributed transparently.
- Farmers are empowered to participate in joint planning, and their knowledge becomes an invaluable contribution to the planning.

- Disciplines and agencies working in an area where decentralised planning is practised need to collaborate in providing services, since the capacity of each is unique.
- Technical officers realise that it is professionally more rewarding to plan activities with the community rather than conduct 'awareness creation' activities designed to persuade farmers to carry out centrally identified activities.

Lessons learned and recommendations

The approach was found to require refinement if it was to become truly participatory, empowering farmers and responding to their needs. These refinements respond directly to difficulties encountered during implementation. Some issues requiring attention are as follows:

- Community feedback should be through village elders and leaders of organised social groups rather than through *barazas* (open public meetings). The latter tend to have a number of drawbacks: inconsistent attendance; lack of specifics, which are not possible in a large meeting; inability to allocate and follow up responsibilities in a large forum; a tendency to attract the less active and under-employed members of the community – hence, those attending *barazas* may not be the most responsible or active persons in the community. Community representation, through genuine elders or group leaders, is instrumental in planning, budgeting, and implementing community development programmes and in monitoring activities and resources. Consideration of gender is imperative in such representation.
- Two key factors caused logistical difficulties in implementing the pilot:
  1 Funding for all participating line ministries was channelled through one government department. The 1995 LLP review suggested that for effectiveness and for the participating ministries to have a sense of full involvement in the planned activities, funds should not be distributed in this manner (Holding *et al.* 1995). If funds were channelled to a district project head, clearly specifying activities and the roles of the various players, this problem would pose fewer challenges. However, all channelling must be complemented by an efficient disbursement and accounting system.

2 As planning did not effectively take into account the already existing commitments of field officers to plans of their line ministries, conflicts of responsibility occurred. Thus, the timing of the joint consultative forum with the various ministries becomes crucial for the harmony of the joint plan with the wider sectoral and national plans of the participating line ministries.

- Technology needs to be developed and trials conducted on representative farms, so that a larger number of farmers are able to observe the relevance and transferability of innovations to their own situation.

- Regular production and timely distribution of reports and minutes of meetings are necessary to maintain involvement of all stakeholders.

- A memorandum of understanding, backed up with a jointly developed work plan, is vital so that all stakeholders are aware of what their roles are and are committed to the process.

- Lack of technical knowledge on particular aspects of the work plan, such as the marketing of farm produce and access to credit, in which extension agencies are generally weak, should be recognised in the early stages, and necessary assistance sought from outside to address them.

- As funding of extension services declines, the responsibility for planning and seeking services and conducting farmer-to-farmer extension activities between villages falls to the rural population. If a community has experience in planning and budgeting, it is better able to take up these challenges.

In summary, LLP was an early attempt to pilot an integrated and participatory extension methodology under the auspices of the Forestry Department. Despite the logistical difficulties indicated, the pilot activity noticeably succeeded in facilitating community planning and implementing a range of natural resource management activities. The communities of Subego and Weru continue their development activities, mobilising their own resources and engaging the services they require. If the pilot is well managed and implemented, the communities remain with a sustainable development agenda and the means to mobilise resources to implement it. For example, a leading national newspaper recently published a double-page spread on how Weru community had organised itself to build a road through the village.

Linking pilots to policy

How does one scale up these pilot activities? How does participatory technology development spread horizontally from one community to the next? How does one institutionalise participatory approaches in extension and research and scale them up vertically?

Our experience has shown which factors are crucial in allowing extension methodology pilot projects to reach their potential:

- consultation at all levels (local, district, and national), before, during, and after implementation;
- a jointly prepared work plan committing all stakeholders;
- relevance to the national and regional context and placement within it;
- the broad participation of a wide range of agencies in implementation;
- a broad ownership of the process, and willingness to allow other partner agencies to develop and adapt the methodology;
- adaptability in the face of a changing policy environment and resulting circumstances and needs of farmers; for example, the current trends towards deregulation and liberalisation would require greater attention being made to forming and supporting local economic organisations in pilot areas;
- measured and factual documentation of the methodology and the results;
- sharing of results in a wide range of forums, with stakeholders;
- honesty about the difficulties and lessons learned;
- provision of constructive suggestions on possible ways forward.

Pilots conducted in an institutional context as part of national debate on extension methodologies and approaches do have an influence on national policy. Pilot project findings can be ingrained in policy development by involving policy-making bodies in their conceptualisation and implementation with a view to influencing their attitudes.

The results of LLP were shared with policy makers in the following forums:

- national agroforestry extension workshop at Masinga Dam in 1995;
- field visits of policy makers to project sites;

- agricultural extension policy team mandated to collect and collate farmers' views; visits made to Subego and Weru;
- donor experience of this project and donor representation at policy meetings.

Policy makers exposed to LLP have been involved in developing the programmes described below.

Building on local-level initiatives

In 1995, the Swedish International Development Cooperation Agency (SIDA), within its ongoing programme support to Kenya, decided to explore further the possibilities of developing methodologies and approaches at the local level that aimed to promote interdisciplinary consultation and collaboration. This was to make it possible for local people to take the lead in their development work, using their own ideas and activities. This approach, building on the experiences of LLP, was aimed at institutionalising multi-disciplinarity and complementarity in all aspects of the rural development process, to achieve synergetic impact of interventions by various development bodies. This process was labelled 'local-level initiatives' (LLIs).

The SIDA-funded programmes implemented by the government of Kenya are operating in the sectors of health, water, public works, and agriculture. It is the management committee, drawn from these sectors, that has developed the LLI approach. The LLI concept has been discussed among representatives of these programmes and joint field trips have been undertaken, in which the concept has been discussed extensively with farmers, farmers' interest groups and institutions, NGOs, government officials, the private sector, and other donor-funded programmes. The committee decided to take on the challenge of developing the LLI concept further and put a pilot into operation in a sublocation of Meru District. At the time of writing, the pilot has been going for 18 months.

Although the project is ongoing, lessons emerging so far from the pilot are already playing an important role in policy development as regards SIDA support to the agricultural sector. The local-level initiative process involved policy makers within the partnership in conceptualising and implementing the project with a view to continually influencing their attitudes. An interactive link between the policy makers and those implementing the project in the field is encouraged in testing this concept. Policy makers are involved in participatory monitoring and evaluation activities. All extension workers involved share in the lessons emerging from the pilot in two annual workshops and quarterly joint field visits to project sites by policy makers and policy implementers.

National Agriculture and Livestock Extension Project, Ministry of Agriculture

Impact analysis of the just-completed National Soil and Water Conservation Programme supported by SIDA has shown that increase in production and productivity has not lifted the population out of the poverty spiral, since the overall income per capita remains less than US$1 per day for the majority of small farmers. Production systems have not adapted to the changing need for a subsistence farmer to move from food security to economic security.

Lessons learned so far from the LLI pilot area indicate that agricultural systems should address the use of higher-value inputs with added value for higher economic outputs. Entrepreneurial farmers who have conserved their land need to be encouraged to diversify their farming systems, incorporating high-value crops. The extension service of the Ministry of Agriculture can help identify long-term markets that small-scale farmers can exploit.

Agroforestry has a contribution to make in meeting these challenges, as demonstrated by the gains that farmers realise through producing such high-value tree crops as fruit, timber, and non-wood products. Multi-purpose trees and shrubs fix nitrogen, control soil erosion, enhance soil fertility, produce fodder that can substitute in a feeding programme for dairy meal (hence saving cash outlay), and perform other on-farm functions.

Emerging needs have been addressed within the framework of the new National Agriculture and Livestock Extension Project (NALEP) (Ministry of Agriculture, Livestock Development and Marketing 1995, 1998, 1999). The experiences of the LLI pilot project were used by the Ministry of Agriculture and Rural Development and SIDA as a basis for formulating the new Swedish support to NALEP under the ongoing agricultural-sector reforms. As the pilot is still underway, it continues to provide NALEP with experience in local-level planning and people's participation. These experiences reflect the wishes of a community in developing a national policy that is aimed at putting in place a pluralistic extension approach. The pilot has shown policy makers that public support is indeed necessary to promote private-sector extension initiatives as well as a strong partnership among the stakeholders.

This new agricultural extension policy will incorporate greater participation in decision making by the various stakeholders in the sector, including farmers, farmer organisations, input suppliers, agro-processors, financial organisations, government, donors, and NGOs. Under the new policy, extension programmes will be based on participatory planning and budgeting with strong emphasis on a bottom-up approach (Nkanata 2000). Forums for beneficiaries and stakeholders will be created for participatory planning and learning. Farmers will be sensitised and trained in legal rights in natural resources management as well as their right to demand transparency and accountability in public extension services.

Note

1 The implementing consultancy firms were Enso Forest Development Oy Ltd and Widagri Consultants Ltd.

References

Antholt, C.H. (1994) *Getting Ready for the Twenty-First Century: Technical Change and Institutional Modernization in Agriculture*, World Bank Technical Paper No. 217, Washington DC: World Bank

Berdeque, J.A. (1998) 'The institutional context of farming systems research-extension in Latin America', in *AFSRE 15th International Symposium Proceedings, Pretoria, South Africa, December 1998*, Pretoria: Association for Farming Systems Research-Extension

Carney, D. (1998) *Changing Public and Private Roles in Agriculture Service Provision*, London: ODI

Ellis, F. (1999) 'Rural livelihood diversity in developing countries: evidence and policy implications', ODI Natural Resources Perspectives No. 40, April, London: ODI

Farrington, J. (1998) 'Organizational roles in farmer participatory research and extension: lessons from the last decade', ODI Natural Resources Perspectives No. 27, January, London: ODI

Franzel, S., C. Holding, J.K. Ndufa, O.C. Obuya, and S.M. Weru (1996) 'Farmers and trees: linking research and extension', *Agroforestry Today* 8(3): 19–21

Holding, C., L.B. Bwana, J.G. Muhoro, S.M. Karanja, J.W. Mathenge, M.A. Mwakuphunza, S.M. Weru, M.N. Kinuthia, and K.K. Kareko (1995) *Local Level Planning in Nyandarua District: A Two-Year Review*, Technical Report No. 15, Nairobi: Nakuru and Nyandarua Intensified Forestry Extension Project, FINNIDA, in co-operation with the Ministry of Environment and Natural Resources, Forest Department

Holding, C. and K.K. Kareko (1997) 'What happens after participatory planning? Participatory implementation: the experience of local level planning in Kenya', *Forest Trees and People Newsletter* 32/33: February

Hoyhta, T., M. Kariuki, P. Njuguna, and K. Wamichwe (1998) *Nakuru and Nyandarua Biomass Survey. Final Report, Kenya–Finland Forestry Programme*, Nairobi: FINNIDA, in

cooperation with the Ministry of
Environment and Natural Resources,
Forest Department

Ministry of Agriculture, Livestock
Development and Marketing (1995)
Agricultural Sector Review 1995,
Nairobi: Agricultural Sector Invest-
ment Programme Secretariat,
Government of Kenya

Ministry of Agriculture, Livestock
Development and Marketing (1998)
*Institutional Arrangements for ASIP
Implementation: Implementation
Manual for the New Structure and
ASIP Management*, Nairobi:
Agricultural Sector Investment
Programme Secretariat, Government
of Kenya

Ministry of Agriculture, Livestock
Development and Marketing
(MOALDM) (1999) *The National
Agricultural and Livestock Extension
Programme (NALEP)*, Nairobi:
MOALDM

Mosse, D. (1994) 'Authority, gender and
knowledge: theoretical reflections on
the practice of participatory rural
appraisal', *Development and Change*
25(3): 497–526

Niemi, T. (1995) *Strengthened Forestry
Extension in Primary Schools: An
Experience of 72 Primary Schools in
Nakuru District, Kenya*, Nakuru and
Nyandarua Intensified Forestry
Extension Project Technical Report
No. 10, Nairobi: FINNIDA, in
cooperation with the Ministry of
Environment and Natural Resources,
Forest Department

Njuguna, P.M., C. Holding, and C.
Munyasa (2000) 'On-farm woody
biomass survey: results of two aerial
and ground surveys, 1993 and 1998',
in *Off-Forest Tree Resources of Africa,
Proceedings of a Workshop, Arusha,
Tanzania, 12–16 July 1999*, Nairobi:
Africa Academy of Sciences

Nkanata, J.N. (2000) 'Briefing note:
divisional extension coordinators
training, Nyeri, May 2000', unpublished
note, Nairobi: MOALDM

Scarborough, Vanessa (ed.) (1996)
Farmer-led Approaches to Extension,
ODI Agricultural Research and
Extension Network, Paper 59a,
London: ODI

Tengnäs, B. (1992) *Guidelines on
Agroforestry Extension Planning in
Kenya*, Nairobi: Regional Soil
Conservation Unit, SIDA

Veldhuizen, Laurens van, Ann Waters-
Bayer, and Henk de Zeeuw (1997)
*Developing Technology with Farmers: A
Trainer's Guide*, London: Zed Books

World Bank (1994) *Adjustment in Africa:
Reforms, Results and the Road Ahead*,
New York: OUP

More effective natural resource management: using democratically elected, decentralised government structures in Uganda

Thomas Raussen, Geoffrey Ebong, and Jimmy Musiime

As forest and plantation reserves decline, the demand for tree products and services steadily increases in the densely populated south-western highlands of Uganda. Farmers are willing to grow trees on their farms but, as is typical in the highlands of Central Africa, on a small farm of less than 1 ha the farmer cannot set aside an area specifically for trees. Integrating trees into the farming system can provide important benefits to the farmer and the environment.

Two types of problem inhibit wider adoption of agroforestry:

- Knowledge and skills about agroforestry innovations are lacking, as are tree seeds and seedlings.
- Some of the problems for which agroforestry is a possible solution must be handled co-operatively by the community rather than by the household (Garrity 2000). This is particularly the case for managing watershed resources in areas with non-consolidated, fragmented farms, which are common in south-western Uganda.

Successful and sustainable community-based approaches to managing watershed resources, of which agroforestry is an important component, share a number of requirements (Cooper and Denning 2000; Garrity 2000):

- Management approaches, as well as the proposed innovations, should be demand driven.
- A set of suitable innovations, such as agroforestry practices, and their key inputs, such as germplasm, needs to be available.
- Efficient community organisations facilitate working together and resolving conflicts.
- Scaling-up efforts need to be co-ordinated and facilitated.
- A 'minimum external input strategy' needs to be put in place.

While farmers and local organisations are quite capable of developing and fine-tuning innovations, they benefit greatly from being exposed to new approaches and technologies. All the other factors listed can be substantially promoted through efficient local governments, as shown in this case study from south-western Uganda.

In this paper, we describe hands-on experience with community-led management of a watershed in Kabale District, and we identify what we consider to be the important components of a successful strategy. We estimate that in Kabale District alone, more than 120,000 km of contour hedgerows will be required for soil conservation. At an average rate of 3000 seedlings per kilometre of hedgerow, this means about 360 million seedlings. The scope of this task makes intervening in the traditional project mode too slow and expensive. We argue that farmers and local government councils must lead jointly in this task if it is to be achieved cost effectively and in reasonable time. Democratic decentralisation of government functions appears to be a key policy factor that is enabling successful watershed management.

The study area

The study was conducted in the 970 ha Katagata watershed in Bubare and Harmurwa Subcounties of Kabale District, which lies approximately between latitudes 1°S and 1°30'S, and longitudes 29°18'E and 30°9'E. The district is mountainous, with altitudes ranging from 1220 to 2500 m (Rwabwoogo 1997). The topography is rugged, characterised by broken mountains, scattered Rift Valley lakes, deeply incised river valleys, steep convex slopes of 10–60°, and gentle slopes of 5–10° adjacent to reclaimed papyrus swamps.

The watershed, in common with about 70 per cent of the land in the district, is covered with ferralitic sandy clay loams (Harrop 1960). Clay loams developed from phyllites predominate on the slopes, while silty clay and peat developed from peaty clay alluvium occur in the valleys. More than 50 years ago, farmers began developing bench terraces along the contours of the hills, and these are now a common feature in Kabale District farming systems.

Kabale District has a temperate climate with bimodal rainfall, averaging 1000–1500 mm annually. Mean maximum and minimum temperatures are 23°C and 10°C, respectively (Department of Meteorology 1997). Although the area is mountainous, the favourable climate and the originally fertile soils coupled with historical factors have led to high population densities of about 246 people per km² (Rwabwoogo 1997).

Smallholder agriculture is based on annual crops of sorghum, bean, and potato. Goats, sheep, and cattle are common, with upcoming dairy production based on fertile pastures at the valley bottoms and zero-grazing units.

The Katagata watershed is typical of the district. It covers 9.7 km² (about 0.5 per cent of the district) and comprises eight villages, two parishes, and two subcounties (see Figure 1).

The policy framework

Local governments have become particularly important in Uganda since the mid-1980s when 'resistance councils' were established to help stabilise the country's security after more than a decade of civil unrest. In 1997, the Local Governments Act of Uganda (Republic of Uganda 1997) initiated an ambitious and much broader decentralisation programme. Government functions were strengthened not only in Uganda's districts, but also at lower administrative levels (see Figure 1). Fiscal responsibility as well as legislative power has been decentralised. For example, the subcounty collects from every adult male a graduated tax and retains 65 per cent of it. The remaining 35 per cent is shared among the county councils (5 per cent), parishes (5 per cent), and village councils (25 per cent). Levies and fees as well as allocations of unconditional and conditional grants from central government add to the budgets of sub-counties and districts. This gives lower levels of administration, beginning with the subcounty, and a quite substantial budget, which may surpass US$100,000 even for a rural subcounty. Equally important is that by retaining much of the local taxes and fees, the local administration becomes directly answerable to its constituency.

The provisions for local government elections guarantee widespread representation at the various councils and include quotas by gender, so that at least one-third of the councillors must be women (see Figure 1).

The Local Governments Act specifies functions and services that a district council can devolve to subcounty councils (LC 3) (Section 31 [4] Local Governments Act, Republic of Uganda 1997). For managing natural resources, these include:

- providing agricultural ancillary field services, such as extension;
- controlling soil erosion and protecting local wetlands;

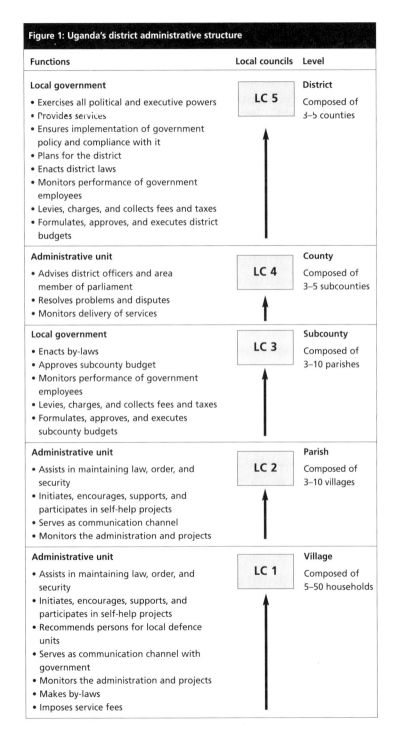

Figure 1: Uganda's district administrative structure

Functions	Local councils	Level
Local government • Exercises all political and executive powers • Provides services • Ensures implementation of government policy and compliance with it • Plans for the district • Enacts district laws • Monitors performance of government employees • Levies, charges, and collects fees and taxes • Formulates, approves, and executes district budgets	**LC 5**	**District** Composed of 3–5 counties
Administrative unit • Advises district officers and area member of parliament • Resolves problems and disputes • Monitors delivery of services	**LC 4**	**County** Composed of 3–5 subcounties
Local government • Enacts by-laws • Approves subcounty budget • Monitors performance of government employees • Levies, charges, and collects fees and taxes • Formulates, approves, and executes subcounty budgets	**LC 3**	**Subcounty** Composed of 3–10 parishes
Administrative unit • Assists in maintaining law, order, and security • Initiates, encourages, supports, and participates in self-help projects • Serves as communication channel • Monitors the administration and projects	**LC 2**	**Parish** Composed of 3–10 villages
Administrative unit • Assists in maintaining law, order, and security • Initiates, encourages, supports, and participates in self-help projects • Recommends persons for local defence units • Serves as communication channel with government • Monitors the administration and projects • Makes by-laws • Imposes service fees	**LC 1**	**Village** Composed of 5–50 households

- taking measures to prohibit, restrict, prevent, regulate or abate destruction of grass, forest, or bush by fire, including the requisition of able-bodied males to extinguish such fires and to cut fire-breaks and generally protect the local environment;
- providing measures to prevent and contain food shortages, including relief work, the provision of seed, and the storage of foodstuffs.

All of these functions and services are relevant to adopting agroforestry innovations in the community. While many councillors are aware of these provisions, they often ask for technical support in order to translate them into action. Others need to be made more aware of the usefulness of a community-based approach as well as the legal backing and the obligations they have. A number of programmes are in place to improve the capability of local councils. All levels of local government have the specific task of advising higher levels of government and can thereby influence policy.

An interesting example of such community action is emerging in Kabale District in south-western Uganda, where farmers in the Katagata river catchment of Bubare and Hamurwa Subcounties (LC 3) have moved forward to begin managing a critical watershed in which soil erosion and related sedimentation are serious problems (Raussen 2000).

Demand-driven approach

A crucial pre-requisite for successful community action appears to be a common understanding that an important problem exists and that communities are willing to invest resources to tackle it. During the exceptionally heavy El Niño rains of 1997–8, farmers of Kyantobi village at the lower end of the Katagata river catchment experienced problems of erosion in the fields on the steep slopes and flooding and sedimentation on their best valley-bottom soils. This erosion during heavy rainfall leads to massive loss of fertile topsoil on the slopes; destruction of crops, particularly at the valley bottoms; and deposits of infertile sand and at times even large stones on the fertile valley-bottom soils. Although the causes of these problems usually lie in the upper parts of a watershed, the immediate impact is highest in the lower parts.

For help to deal with the problems, representatives from the village at the lower end of the watershed contacted the Agroforestry Research and Development Project jointly implemented by the Forestry

Resources Research Institute and the International Centre for Research in Agroforestry (ICRAF).

When agroforestry dissemination staff visited the watershed, it became obvious to project staff and farmers alike that any effective measures to help control the problems of runoff would require community action throughout the watershed. This is particularly so since farmers' fields are fragmented, and erosion control in a single field on a given slope would not have any significant effect. Project staff made it clear that the project could help with training and materials for soil conservation but that local leaders would have to organise the key element for success – community action. The villagers readily accepted this condition. Community arrangements are already common for grazing regulations to protect crops and planted trees and to prevent fires. Community action is possible because local authorities can make suitable arrangements and village and subcounty councils can set and implement by-laws. Furthermore, local governments can help identify community needs and organise discussions on possible solutions. Higher levels of local government assist lower levels to initiate contacts with relevant organisations that may help in implementing projects. This means that a system is in place for bottom-up planning of projects.

Available agroforestry innovations

Obviously, community action requires suitable innovations. Often these may be available locally and may only require modification. However, research institutes can often provide further inputs to this process by:

- advising farmers on how to set up tests to explore the best adaptation of the innovations; and by

- introducing innovations to the farmers. For example, in the study area it is common to leave strips of natural vegetation at the terrace risers. However, these strips are not sufficiently stable to withstand the impact of runoff during heavy rainfall. Through their village council (LC 1), the Kyantobi farmers selected delegates who were taken by the project staff on a one-day study tour to on-station and on-farm research sites. This exposure led them immediately to identify contour hedgerows as the most suitable innovation for alleviating the erosion problems. These hedges provide not only adequate soil and water conservation services (Cooper *et al.* 1996) but also products such as high-quality dairy fodder, stakes for climbing beans, and fuelwood.

Usually, more than one best-bet innovation is required for farmers to experiment with and adapt. These may be specific to various farming conditions in the area. This possibility of trying things out for themselves is also necessary to keep farmers' enthusiasm high.

How do farmers find agroforestry innovations? One of the most successful means for giving farmers informed choice and to promote innovations is exchange tours to visit farmers already using them. Here again, local councils can be helpful in choosing suitable participants, making logistical arrangements, and perhaps covering part of the costs such as by providing transport or food. During such tours, farmers from Kyantobi village identified other agroforestry innovations that they also wanted to try on their farms:

- *Boundary planting* with upperstorey trees (for example, *Grevillea robusta* and *Alnus acuminata*) to produce much-required poles and timber without foregoing much of the productive cropland.

- *Rotational woodlots* on degraded land for fuelwood and stake production while at the same time improving the soil (AFRENA–Uganda 2000).

- *Fruit trees* for home consumption of fruit, particularly the newly introduced deciduous trees (apple, pear, plum), which can produce crops in the highlands and generate cash in urban markets in lower-lying areas (AFRENA–Uganda 2000).

Community organisations

Why are effective community organisations so important for disseminating agroforestry information and systems cost-effectively and successfully?

Most dissemination about agroforestry is currently done in a project mode. Much effort is required to establish suitable structures for the process, which may include forming dissemination groups, resolving agroforestry-related disputes in the community, and posting extension officers in the target areas.

Working through established community groups allows the development organisation to concentrate on what it is best at: providing training and the few necessary materials. It also allows the local council to concentrate on its strengths: planning, mobilising the community, facilitating joint efforts, and resolving conflicts. These functions are important, particularly if one considers how much time and funds development organisations, as outsiders, usually invest to provide these services. Democratically elected village or parish councillors are respected and well-placed to fulfil these functions more cost-effectively.

Local communities, as they plan, often benefit from the technical backup that development organisations can provide. In our case study, for example, villagers much appreciated the participatory mapping exercise, both for its team building and for its usefulness as a tool for planning natural resource management. Farmers met in the field and mapped a whole slope. To their own surprise, it was not always easy to identify the owners of fields (over 40 on one slope). They then determined the measures required for soil conservation. Based on their map (see Figure 2), the dissemination staff found it simple to calculate the length of the contour hedges and the number of seedlings that each farmer would need (see Table 1). This approach is an important improvement over the common practice in which projects determine a rather abstract target for nursery production (often based on donor rather than farmer demand). In the case study, each farmer could now decide the number of seasons required to raise the seedlings and whether to do this individually or with a group of fellow farmers. We expect this approach to have a strong motivational effect on the farmers.

Empowering farmer groups and their local councils to plan and implement the conservation exercises should enhance the scaling-up process. Already in the Katagata watershed, 164 farmers have become involved in agroforestry and have established 32 nurseries. As mentioned, several hundred million seedlings would be required to establish contour hedges all over Kabale District. This can be achieved in a reasonable time only if planning, raising seedlings, and establishing them in the field becomes a self-propelled and sustainable exercise. Local councils appear to have the authority and most of the resources to lead this process.

Local government

Importantly, local government in villages and parishes can instigate community action and resolve conflicts; higher levels in the hierarchy have their strengths in co-ordinating, making contacts and requests, assisting in monitoring, and providing funds.

In Uganda, a typical district contains between 15 and 20 subcounties, and the subcounty appears to be the suitable unit for undertaking these

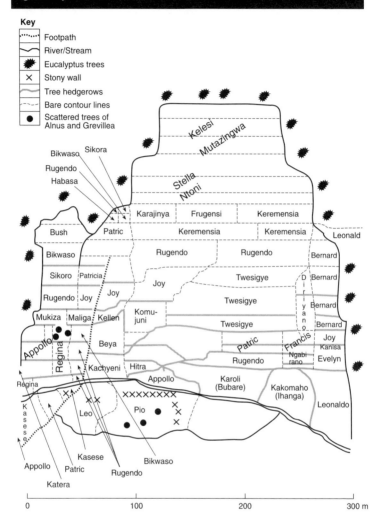

Figure 2: Kyantobi watershed area

Key

- ········· Footpath
- ∿ River/Stream
- ✸ Eucalyptus trees
- × Stony wall
- ≈ Tree hedgerows
- --- Bare contour lines
- ● Scattered trees of Alnus and Grevillea

functions. It is the lowest level with budgetary power and corporate rights, as it keeps and spends 65 per cent of the graduated tax it collects. This implies that the subcounty administration is directly responsible – and answerable – for using the main tax of its constituency. Subcounty leaders are in direct contact with all their electorate and will in most cases work towards re-election by providing good services. While agroforestry or natural resource management is probably not the highest priority (health and education usually are), the farmers who

Table 1: Exercise in participatory planning of soil conservation at landscape level

Name	Hedge length (m)	Seedlings (no.)	Amount of seed (g)	Source of seedlings
Rhoda Mukiza	60	240	36	Own nursery
Apollo Oworinawe	135	540	81	Own nursery
Regina Turyatemba	70	280	42	Own nursery
Biramahire John	30	120	18	Group (3) nursery
William Rugendo (Kyantobi)	310	1240	186	Group (1) nursery
Sikora Karya Gokwe (Katungu)	70	280	42	Own nursery
Bikwaso Thomas	Woodlot	0	0	Own nursery
Kellen Apuchi	60	240	36	Group (3) nursery
Joy Mbahunami	195	780	117	Group (3) nursery
Patric Turyahikayo	220	880	132	Group (3) nursery
Beya Rubereto	90	360	54	Own nursery
Kacyeni (Kashakyi)	60	240	36	Group (1) nursery
Habasa T	30	120	18	Own nursery
Kayinya John	75	300	45	Own nursery
Hiltra Micheal (Katungu)	100	400	60	Own nursery
Komujuni James	70	280	42	Own nursery
Barijunakyi Adonia (Kyantobi)	90	360	54	Group (2) nursery
Frugensi Butamanya (Karubanda)	105	420	63	Group (3) nursery
Keremensia Birigo	135	540	81	Own nursery
Twesigye Justus (Kyantobi)	380	1520	228	Own nursery
Diriyano Ziranga	60	240	36	Group (3) nursery
Twesigye Francis	85	340	51	Own nursery
Ngabirano Vicent	125	500	75	Group (1) nursery
Karori Nyakana	145	580	87	Group (1) nursery
Bernard Karimarwakyi (Kyantobi)	110	440	66	Group (1) nursery
Kariisa Benoni	60	240	36	Group (1) nursery
Evelyn Tibemanya	60	240	36	Own nursery
Kakomaho Peter	75	300	45	Own nursery
Leo Nkirirelhe (Mwiguriro)	70	280	42	Own nursery
Gerera	0	0	0	Own nursery
Kelesi Mutazingwa (Ihanga)	420	1680	252	Group (2) nursery
Stella Ntoni (Ihanga)	280	1120	288	Own nursery
Pio (Kashakyi)	175	700	105	Group (2) nursery
Kasese	150	600	90	Own nursery
Total number of seedlings	4300	17,200	2580	

Using the map (see Figure 2), farmers estimate the length of the contour hedges they will need; assuming four seedlings per metre of hedge, they then determine the quantity of seedlings and seeds.

depend on the sustained productivity of their land will appreciate a leader's efforts in this direction.

It is essential for the research and development organisations to participate in local government planning processes, because only then will local governments perceive them as true partners. Together they can design sustainable development plans. It proved helpful for the subcounty administration and the Agroforestry Research and Development Project to sign a joint memorandum of understanding. While this only spelled out the broad basis of the collaboration the document positively influenced the perception of co-operation on both sides. For example, it became common for project staff to be invited to all environment-related meetings, and it was accepted that agroforestry activities would become part of the subcounty's workplan, including budget allocation for them. The project, in its turn, supported such activities as typesetting and producing a quarterly subcounty bulletin, although only a small part of it referred to environmental issues.

An initiative to explore would be for local governments with similar problems or programmes to establish ways to network so they could share resources and information, such as on natural resource management.

Minimum-input strategies

Large-scale agroforestry adoption has to be affordable if it is to be successful and sustainable. Firstly, the innovations themselves should require minimum inputs in terms of labour and cash. Secondly, if agroforestry is to be adopted on a wide scale, the dissemination approaches need to be low-cost. This is even more the case if the main inputs are not expected from development organisations but from communities themselves. Scaling up agroforestry, which will largely have to be paid for through local people's work and taxes, has to be as cheap as possible in order to be accepted. People's labour and tax funds have to cover a wide range of other communal necessities, which include other and often higher priorities like schools, health, transport, and marketing.

If it is agreed that the project mode is too expensive for widespread, locally supported scaling up of agroforestry, then the fundamental question for any agroforestry extension programme becomes: can it still be successful with less labour and fewer inputs than are generally available in projects?

Developing these innovations in a research and development continuum (ICRAF 2000) should involve farmers at all stages, allowing them to simplify the methods. If local government structures are to lead the dissemination process, only the absolute minimum of required inputs should be externally provided – in most cases, training and germplasm. Most farm nurseries will not require polythene tubing, wheelbarrows, shovels, rakes, and watering cans. Farmers have for decades raised their vegetable seedlings without these items and can raise many of their tree seedlings the same way. It is, however, acknowledged that quality fruit trees, for example, require a different approach, since they need higher inputs and probably specialised commercial nurseries.

Similarly, the momentum and widespread adoption of agroforestry innovations will depend on whether local councillors will facilitate the adoption as part of their regular duties and not as an additional 'project' service that needs to be paid for externally. It is therefore important that councillors be trained and become aware of the programmes, so that their perception of environmental issues and interventions is raised and their willingness to allocate their constituency's resources towards such issues is increased.

The true decentralisation process in Uganda is only two years old. So we are seeing just the beginning and have a unique opportunity to learn from it. We should begin to develop clear development hypotheses in relation to the perceived potential of local community action and local government structures. Only then will we be able to test them and discover whether the potential really exists.

For research and development organisations, the opportunities are tremendous. We could test innovations with hundreds or thousands of farmers, explore their impact on watershed and landscape scales, and monitor farmers' modifications. Monitoring is particularly important since widespread testing by farmers, coupled with a workable monitoring system, may initiate a true evolution. When thousands of farmers undertake small trial-and-error experiments, we can expect the 'fittest' innovations to survive.

Conclusions

Scaling up adoption of agroforestry innovations from individual farms to watersheds and whole farming systems is a formidable task. Despite the impressive impact made by various agroforestry development projects in south-western Uganda, the task is far too large to be accomplished in a project mode. Only if communities – convinced by the success of early agroforestry adopters – take responsibility for searching for solutions, adapting and adopting them to their complex environmental problems, and implementing them on a large scale, will environmental degradation in the watershed be addressed in time and with affordable resources. This requires enabling the community to understand the problems and plan interventions.

Local farmer organisations and local governments are best able to mobilise the community and solve local problems, with research and development organisations providing technical backup and quality germplasm. This proposed mode is different from the traditional technology-transfer approach, in which researchers generate technologies and extension specialists extend them to farmers. Here we propose enabling farmers to analyse and plan a range of options and solutions. Most importantly, they should themselves identify these options and solutions and maintain an open and regular dialogue with all the institutions involved. Another key ingredient for a successful approach is patience: patience to allow initiatives to grow and farmers to plan and explore them for themselves.

The scaling-up process this paper describes is still in its infancy and needs more social research and quantification. However, the achievements made with limited physical inputs from outside are remarkable.

Uganda is advanced in the decentralisation process; however, even in countries with weaker local governments, the potential to make use of local organisations in scaling up innovations often appears to be untapped. Greater efforts are needed to mobilise local government officials as promoters of natural resource management practices.

References

AFRENA–Uganda (2000) *Agroforestry Trends,* Kampala: Agroforestry Research Network for Africa

Cooper, P.J.M. and G.L. Denning (eds) (2000) *Scaling Up the Impact of Agroforestry Research,* Nairobi: International Centre for Research in Agroforestry

Cooper, P.J., R.R.B. Leakey, M.R. Rao, and L. Reynolds (1996) 'Agroforestry and the mitigation of land degradation in the humid and sub-humid tropics of Africa', *Experimental Agriculture* 32: 235–90

Department of Meteorology (1997) *Monthly and Annual Weather Data for Kabale District,* Kampala: Uganda Ministry of Natural Resources

Garrity, D. (2000) 'The farmer-driven Landcare movement: an institutional innovation with implications for extension and research', in P.J.M.

Cooper and G.L. Denning (eds) *Scaling Up the Impact of Agroforestry Research*, Nairobi: International Centre for Research in Agroforestry

Harrop, J.F. (1960) *The Soils of the Western Province of Uganda*, Memoirs of the Research Division, 1/6, Kampala: Uganda Department of Agriculture

ICRAF (2000) *Paths to Prosperity through Agroforestry: ICRAF's Corporate Strategy 2001–2010*, Nairobi: International Centre for Research in Agroforestry

Raussen, T. (2000) 'Scaling up agroforestry adoption: what role for democratically elected and decentralized government structures in Uganda?', in P.J.M. Cooper and G.L. Denning (eds.) *Scaling Up the Impact of Agroforestry Research*, Nairobi: International Centre for Research in Agroforestry

Republic of Uganda (1997) *The Local Governments Act*, Acts Supplement No. 1 to the *Uganda Gazette* 19(40)

Rwabwoogo, M.O. (ed.) (1997) *Uganda District Handbook*, Kampala: Fountain Publishers

On-farm testing and dissemination of agroforestry among slash-and-burn farmers in Nagaland, India

Merle D. Faminow, K.K. Klein, and Project Operations Unit

The State of Nagaland is located in north-eastern India along the border with Myanmar (Burma). It is almost entirely populated by indigenous people belonging to at least 17 different tribes (see Figure 1). Agriculture in Nagaland is primarily oriented towards subsistence, and the swidden systems that predominate in most areas of the state are the main cause of the rapid deforestation that has occurred in the last 30 years. Rapid population growth, limited opportunity for sedentary agriculture, mountainous topography, and the wealth of native tree biodiversity in Nagaland have led some development specialists familiar with the region to propose agroforestry as a means of modifying the traditional agricultural production systems and reducing deforestation. When fast-growing high-quality timber trees are integrated into the traditional practice of crop production (usually a two-year cycle, based on upland rice), the subsequent fallow forest cover is enriched and can generate increased revenue when the trees are cut before the next cropping period, 10–15 years later. Thus, income from tree production can eventually become a significant cash source, which would encourage farmers to practise intensified agroforestry instead of extensive swidden. However, experience with agroforestry in other tropical regions indicates only limited success in getting traditional farmers to adopt agroforestry in place of their traditional subsistence systems, and only minimal success in slowing down deforestation.

This paper describes the structure and impact of a development project in Nagaland, which was designed to encourage traditional swidden farmers to adopt agroforestry. We describe and assess how the project was implemented and we estimate its reach and impact. Initial evaluation suggests that agroforestry has spread rapidly and been primarily adopted on land that otherwise would have been used only

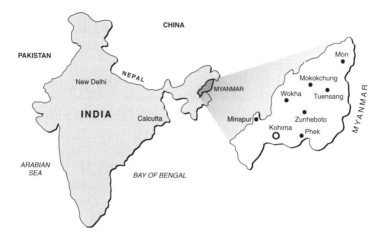

for swidden agriculture. Thus, Nagaland appears to be on a path to intensified land use based on agroforestry, which is likely to reduce the rate of deforestation. Following this, the factors that have affected the extent to which agroforestry has been scaled up are summarised, and lessons are drawn from the experience in Nagaland.

Land-use systems

Farmers in Nagaland practise a form of swidden agriculture called *jhum* in local dialects. As in other swidden systems (Faminow 1998; Sánchez 1976), Nagaland's farmers slash and burn a small plot of forest (often 1 ha or less), plant a wide mix of crops for one or two years of production, and let the land go to fallow, which then usually permits the forest to return eventually. The environmental impact made by each farm family is small, but the cumulative effect on local and regional ecosystems can be substantial. In Nagaland only about 1700 km² of the land is suitable for terracing and irrigation. An area of 7000 km² is subjected to shifting cultivation, 500 km² of which is slashed and burned each year.

With long periods of fallow (*jhum* cycles of 15–20 years), shifting cultivation in Nagaland can be sustainable (Ramakrishnan 1993). However, a high birth rate combined with economic stagnation and limited off-farm opportunities has caused the rural population to grow at an annual rate of nearly 4 per cent. About 70 per cent of the roughly

1.5 million population lives in rural villages and is dependent upon agriculture for a livelihood. In some villages in Nagaland, land is now in such short supply that *jhum* cycles have shortened dramatically, and farmers must return to the same plot of land after only three to five years. Yields are plummeting and rapid deforestation of mountaintops has seriously affected water supplies.

Nagaland is a complex mixture of similarities and contrasts. Virtually the entire state is mountainous, and swidden agriculture can be found at altitudes between 300 m to over 1800 m above sea level. In some highland areas of the state (for example, in Kohima and Phek Districts) population pressure is still moderate and ample water supplies allow for terraced irrigation for producing rice, the staple food. There, shifting cultivation is quite limited and is often used for growing animal feeds like millet and maize. In other areas (for example, Mokokchung and Wokha Districts) altitudes are generally lower, population pressure is substantial, and water limitations effectively preclude terraced rice production. Upland rice is generally planted in extensive swidden plantations and the fallow period is often short. Broadly, the swidden systems used in Nagaland appear similar, but substantial variation can often be seen upon closer inspection.

Some of the variation is related to ethnic differences. Although there is still academic debate about the actual number of tribes and languages present in the state, the Naga people can be divided into at least 17 major tribes, each with a different language. Within a given tribal language may exist local dialects that differ substantially from each other. Land-tenure arrangements can vary considerably by tribe and even among villages within a tribe. Private ownership is the norm across Nagaland, but there are variations in traditional property law ranging from individual ownership through to family or clan land custody (which can be passed down to heirs, but not sold). Among two tribes, the Konyak and the Sema, occupying Mon and Zunheboto Districts, village chieftain systems entail that the village chief is the ultimate owner of all village land. Regardless of the land-tenure system in use, however, land is almost always controlled locally, with state ownership limited to about 9 per cent. This contrasts with the communal ownership of natural resources common in other tribal areas of India and Nepal, which are often complicated by unsatisfactory legal frameworks, making the legal basis for common property ambiguous (Arnold 1998). Beckly (n.d.) has pointed to the importance of tenure issues in improving forest management and believes they have not been given adequate attention in studies related to agroforestry.

In traditional Naga cultures, governance in local villages was (and still is) the foundation of society. Historically, although there is some variation between tribes and between villages, villages operated independently and were governed in a way akin to ancient Greek city-states, with a governing council comprising elders with administrative authority (Keitzar 1998). Much of this tradition continues today because, when Nagaland became a state of India, special consideration was taken to ensure that traditional tribal law and practices of land use and ownership continued to take precedence. Villagers meet with their village council before the start of each planting season to plan and coordinate their strategy. In most cases, swidden plots for all families in a village are located together and major activities, such as burning, are co-ordinated. As a result, development projects implemented in the village, with significant decision-making functions residing with local authorities, may offer greater potential for adoption and sustainable change in some circumstances than do top-down strategies.

Identification of solutions

New and improved technologies are sometimes touted as primary weapons in the battle against deforestation, particularly in tropical moist forest areas like Nagaland. Technology can encourage intensification, increase yields, and set the stage for permanent cultivation. Many governments have embraced this idea and actively tried to wean swidden agriculturalists from their traditional practices. However, when top-down solutions are imposed without due consideration of local needs and conditions, they can be ineffective or even damaging. For example, in remote hill areas like Nagaland, efforts to introduce high-yielding rice varieties and production methods must take into account the limited potential for terracing and irrigation, and recognise that many high-yielding varieties of rice that were developed for plains areas do not perform well at altitude. Combined with few marketing opportunities, a lack of commercial tradition in rice, and poorly developed infrastructure, this type of intervention may even create greater food insecurity.

Even when 'hi-tech' solutions appear to be in farmers' interests, indigenous peoples are often vulnerable when interaction with the outside world is increased (Egneus 1990; Higgins 1998; Jodha 2000). They may lack information and experience in cash market participation. 'Participatory' strategies that are culturally sensitive and conform to ecosystem properties so that they modify, rather than replacing, existing systems might do better (Baker 1998).

Immediately after Nagaland became a state in 1963, the Indian government and the government of Nagaland waged a long and ultimately futile battle to put a halt to jhum cultivation and initiate sedentary agriculture based on intensified annual cropping (NEPED and IIRR 1999). Nagaland has potential for more irrigation but, because of topographical, market, and traditional cultural constraints, it has proved impractical for most villagers. Ultimately, the government of Nagaland concluded that the *jhum* cultivation system could not be eliminated, at least in the short to medium term.

However, some community groups and Nagaland government officials began to develop locally based solutions to counteract expanding *jhum* and deforestation. During the 1980s, one village in Mokokchung District held a seminar to discuss the problem that they and most other villages were facing: population was increasing, there was little or no opportunity to irrigate, and the *jhum* cycle was decreasing (that is, they had to shorten their fallow periods to grow enough food for subsistence). In areas of extensive swidden, entire mountainsides were being slashed and burned each year. They wanted a solution.

Land-shaping using contour bunds was initially encouraged as the primary corrective measure for soil erosion and declining soil fertility. This concept of land-shaping is a direct derivative of the practice of intensified swidden that one specific group in Nagaland, the Angami, uses, particularly in Khonoma Village in Kohima District. In this intensified swidden system, stone-reinforced terraces are built, and nitrogen-fixing alder (*Alnus nepalensis*) trees planted along the contour to provide firewood and enrich the soil. The yields are high for crops planted in this intensified system and the fallow period is minimal. Typically, two years of crops are planted, followed by only two years of fallow. The resulting highly productive four-year swidden cycle has been maintained in Khonoma for hundreds of years with stable productivity and without any apparent degradation (NEPED and IIRR 1999).

As a result of the seminar held in Mokokchung, two small-scale land-shaping and tree-planting projects were developed to deal with the problems of soil erosion and decreasing fertility. Government officials wanted to undertake projects on a larger scale but felt the projects would not work unless the villagers themselves had designed them (NEPED and IIRR 1999). They explicitly recognised the importance of including grassroots participation – especially the rural poor – in the earliest consultations before drawing up priorities for a larger scale project. The problem is complicated further in Nagaland because jurisdiction over agroforestry extends across several government departments.

It was eventually decided that the development of permanent agroforestry could be a long-term corrective measure for soil degradation, biodiversity management, and income growth, providing a pathway to an improved *jhum* system (Ramakrishnan 1993). Importantly, however, as Mallik and Rahman (1994) have noted, community forestry often meets the basic needs of the community, in addition to market-oriented objectives. Ultimately, it was decided to implement a participatory forest management approach following eight of the practical steps to system improvement that Ramakrishnan (1993, 1996) had proposed:

1 Encourage technology exchange between various ethnic groups.

2 Where possible, maintain a swidden cycle of at least 10 years.

3 Incorporate ecological insights into tree architecture.

4 Introduce nitrogen-fixing trees into the system, such as alder.

5 Maintain important bamboo species to conserve soil and serve as windbreaks.

6 Introduce fast-growing native shrubs and trees.

7 Condense the timespan of forest succession and accelerate recovery by adjusting species mixes in time and space.

8 Strengthen conservation measures based upon the traditional knowledge and value system.

The Nagaland Environmental Protection and Economic Development (NEPED) Project was initiated in 1994 with funding from the India-Canada Environment Facility (a funding mechanism of the Canadian International Development Agency, CIDA) and Canada's International Development Research Centre (IDRC). The project is a large-scale experiment in participatory development emphasising local technology. The strategy for technological development is farmer-led testing, where farmers themselves select agroforestry technologies, implement the field tests, and assume responsibility for disseminating the results locally.

Design of the Nagaland Environmental Protection and Economic Development Project

The goal of NEPED was sustainable management of the natural resource base in land used for shifting cultivation. Objectives were (1) to stimulate interest in agroforestry so that farmers plant trees along with food crops in *jhum* fields; (2) to improve upon soil and water conservation methods; and (3) to encourage and develop local capacity for starting and carrying out development initiatives.

A key factor for successful forestry projects is to adopt the best possible forest technologies, including suitability of species chosen for projected end-uses; this adoption should be accompanied by good nursery practices and post-planting tree management (Tamale *et al.* 1995). A top-down approach, which requires significant technical direction and control, was rejected early in the planning stages for NEPED. Instead, NEPED opted for a 'search and find' philosophy that would encourage farmers to experiment themselves, where the project implementation team would provide basic technical advice but adopt more of a facilitation role, especially in disseminating indigenous solutions and strategies.

The principal activity of the project was to establish test plots for farmer-led experimentation and dissemination of agroforestry in each participating village, with the target to establish test plots in every registered village in the entire state. The primary budget outlay was to promote planting trees for enriched fallow in *jhum* fields, along with shaping the land to improve soil management. Each village has a village development board, which is an autonomous, locally directed body that can disburse state development funds. These boards were each asked to select two farmers or groups of farmers to be allocated test plots, one at an upper altitude and another at a lower altitude. In most cases, test plots were allocated to individuals or small informal collectives of three to five farming families. The selection criteria were that the trees must be planted in *jhum* fields scheduled for planting in the year selected and that the farmers chosen should be progressive. Aside from this, discretion of selection was given to the village development boards. Organised larger groups, such as women's societies, church groups, and youth groups, were also allocated test plots if the local authorities selected them. Usually these groups used village-owned land or developed sharing arrangements with individual landowners. Farmers selected for participation received cash payments to offset field-testing costs. Success in farm forestry is generally contingent on enough free seedlings being available (Chatterjee 1995), so the cost of planting material was included. Farmers were then responsible for carrying out the field testing and reporting the results.

Test plots were normally 3 ha but occasionally were larger when villages contributed additional funds or in-kind inputs. Because family-sized plots are typically smaller than 3 ha, most test plots involved small, informal collectives of farm families. The project paid farmers for planting trees and for digging contour trenches (that is, land shaping using bunds). Initially, total payments to farmers amounted to US$245 per hectare, with land-shaping as a significant component (about 75 per cent) of the amount. However, farmers did not readily accept the technique of land shaping with contour bunds; therefore, beginning in 1998 test plot owners were encouraged to use a modified land-shaping technology (traditional erosion control with very small trenches). Total payments were lowered to US$230 per hectare and the land-shaping share to 50 per cent of the total. Payments occurred in three steps and were subject to project officers verifying the work. The initial target was to establish 2000 test plots (two in each of the 1000 villages in Nagaland). This ambitious target to extend the project across all villages was nearly fully achieved.

The agroforestry part of the project was to ensure that farmers planted trees during the first year of a two-year swidden crop sequence, so that food crops were integrated with timber trees. After the second year, fields are normally left fallow, and tree planting would result in an improved fallow as vegetation spontaneously grew again around the planted trees. However, in practice, test plots were generally maintained and managed as tree woodlots after crop harvest. In addition, roughly 10–15 per cent of the test plots were implemented as farm forestry in the low altitudes (under 500 m) where irrigated crop production is ample and swidden agriculture limited. Across the entire state, over 1800 test plots were handled in more than 850 villages.

Beginning in 1996, a gender component was added to the initial management plan, requiring that some test plots be allocated to women's groups. Women were coming forward and asking to be included, even though traditional law in almost all villages does not permit women to own farmland. In total, 93 test plots were allocated to women's groups. Sharing arrangements with landowners were chiefly verbal, and in some cases women only had limited ownership benefits of the trees that were the product of their work.

On the test plots, farmers could experiment with different techniques and systems that they chose. Training and advice to farmers focused on technical aspects related to constructing bunds and on recommendations about tree planting (spacing, species selection, planting method). Farmers, however, were free to select species composition, planting density, and location as long as the area they worked totalled three contiguous hectares. They undertook soil conservation measures, and otherwise maintained the area well. A broad range of tree-planting approaches was allowed and the farmers actually used this range. For example, although project officers suggested relatively sparse tree planting (600 seedlings per hectare), based upon silviculture recommendations to optimise timber yield, most farmers chose much denser planting (often up to 2000 seedlings per hectare). A forestry expert criticised this practice during the NEPED mid-term review. However, interviews with farmers revealed underlying benefits that suited farm livelihood needs that would have been missed if the farmers had been forced to adopt the silviculture-based recommendation:

- Dense planting encouraged straighter trunk development and increased final timber value.

- The upper canopy closed sooner and lowered weeding costs in the early years of tree growth (labour is generally the constraining resource for farmers in Nagaland).

- Available planting material was often of variable quality, and farmers were unable to be selective when establishing plantations.

- Trees exhibiting substandard growth could be harvested after five to seven years of growth and sold in the market for construction poles, thereby providing cash in the intermediate term and allowing remaining trees to achieve better girth.

Over the course of the project, other local practices, ecological insights, and grassroots innovations were observed and disseminated:

- To improve biodiversity of *jhum* lands, planting a range of local species was encouraged. Initially about 100 species of fast-growing local trees with good timber quality were identified. Farmers were trained to use locally available planting material (wild seeds and seedlings) and encouraged to select trees to plant from these local species. Ultimately, roughly 40 species were adopted on a fairly broad scale across Nagaland.

- In some regions of Nagaland, farmers use indigenous methods of erosion control, forming runoff blockades with available materials (split bamboo, logs, rocks). These blockades are effective because they do not disturb the soil integrity. Farmers who used this method did not want to shape their land with contour trenches because the labour cost was three times higher, and erosion control only marginally improved. Cultural resistance prevented introducing other techniques such as contour hedges.

- Some of the local tree species were unknown by forest experts or by the market, but formed part of Naga local botanical knowledge. Persons with expert local knowledge contracted in the course of the project were instrumental in identifying indigenous species with desired characteristics, such as being fast growing, providing good timber quality for various uses, usefulness for firewood, and helped identify the most effective propagation techniques.

- Early on, project officers encouraged planting a diversity of trees in fields. But as a consequence of participatory learning by both farmers and field officers, farmers planted a narrower spectrum of varieties in later years of the project.

- The species planted in test plots included a diverse mixture of mostly local species (see Tables 1 and 2). *Gmelina arborea* was by far the most common species, followed by *Alnus nepalensis* (the nitrogen-fixing alder), *Tectona grandis*, and *Melia composita*. Of the top ten species, only *T. grandis*, teak, can truly be termed an exotic. Some unidentified local species were also planted.

- Forest enrichment through selective weeding became a more prominent management technique, where valuable species were allowed to rejuvenate naturally by being spared while weeding. This helped ensure that test plots retained natural biodiversity and did not rely entirely upon tree seedlings from nurseries.

Scaling up the project

A key indicator of success of this participatory project was the rate at which other farmers in the villages actually replicated the agroforestry strategy by planting trees in their swidden fields and using improved soil conservation techniques, without receiving direct project support. Test plots served as local entry points for testing and disseminating improved technologies for use in *jhum land*. At the outset of NEPED, it

Table 1: Most commonly planted species under NEPED, 1995–7

Rank	Species	No. planted (thousands)
1	Gmelina arborea	2544.0
2	Alnus nepalensis	947.3
3	Tectona grandis	642.4
4	Melia composita	571.0
5	Terminalia myriocarpa	362.0
6	Cedrella spp.	290.0
7	Spondias axilaris	274.5
8	Aquilaria agallocha	204.4
9	Duabanga grandiflora	114.1
10	Unidentified local species	107.8

Table 2: Diversity in trees planted in 1997 test plots, by district

District	No. of species observed	Most common tree species observed
Kohima	17	Gmelina arborea, Tectona grandis
Mokokchung	18	Gmelina arborea, Aquilaria agallocha, Duabanga grandiflora, Tectona grandis
Phek	10	Gmelina arborea
Wokha	10	Gmelina arborea, Tectona grandis
Zunheboto	8	Gmelina arborea, Alnus nepalensis
Tuensang	15	Gmelina arborea, Alnus nepalensis, Melia composita
Mon	11	Gmelina arborea, unidentified local species

was expected that tree planting would scale up rapidly because the plan was that all 1000 villages in Nagaland would participate in the experiment. Land-use decisions in Nagaland are generally taken jointly within a village, so it was anticipated that experiments that proved successful would be quickly adopted by other villagers, who would initially adopt a 'watch and see' strategy. Wide dissemination was also encouraged because farmers were trained on site and non-participating farmers invited to attend training sessions. Follow-up visits by the implementation team to verify progress and provide supplementary training ensured that the project had high visibility in the communities. An important component of this capacity for follow-up visits was that all project officers were assigned a vehicle.

Within two years after the project was implemented, evidence of extensive tree planting became apparent and could be observed casually across the state. Initially, most scaling up was spontaneous, but targeted government programmes later reinforced it. Some of the Nagaland government departments began distributing tree seedlings on a limited scale in 1997–1998. Three districts were not included in the government programme, and seedlings were disbursed to only 150 villages. Then Nagaland declared 1999 as the 'Year of Tree Plantation' and, in a highly visible and publicised initiative, distributed close to ten million seedlings to individuals and organisations across the entire state. The untargeted approach of the government programme in 1999 appeared effective in getting seedlings out into the countryside, but it had significant regional gaps, and often seedlings did not reach lower income farmers.

Survey to measure extent of scaling up

With a modest budget and limited time, an attempt was made to estimate the impact that the NEPED Project made on reforestation and on improving agroforestry techniques among villagers across the state. As data were collected before the Nagaland Year of Tree Plantation programme, they should be considered conservative estimates of the extent to which agroforestry was scaled up. A random sample of villages was selected from among all those that had been awarded test plots in the project. Questionnaires were developed and administered to villagers in the selected communities to gauge the level of participation in improved agroforestry techniques.

Because travel to the remote villages in the study population was difficult, it was necessary to establish upper limits on questionnaire size and sample size. Each of the seven hill districts in Nagaland has subdistricts. In a random draw, two subdistricts were selected from each district. All test plots in the selected subdistricts were eligible for selection, except for a small number that were not easily accessible. Two test plots were then randomly drawn from each of the chosen subdistricts, providing a random sample of 28 test plots. Three separate questionnaires were developed: one for the village council, one for the owner(s) or operator(s) of the selected test plot, and one for the owner(s) or operator(s) of a replicate plot, drawn randomly from lists of replicators provided by village councils. It was sometimes difficult to differentiate clearly between actual project replicators and farmers who would have planted trees anyway, either being influenced by other programmes or for private reasons. However, the other programmes that distributed planting material were initially stimulated and highly influenced by NEPED, so including replicators that these programmes supported does not seriously undermine our evaluation of the project's impact. Before NEPED began, tree planting had been sporadic and limited, so the widespread and extensive spread of the new approach (as documented below) could only be directly due to NEPED or to other government programmes (the primary one, the Year of Tree Plantation, had not yet taken effect when the survey was completed) strongly influenced by NEPED, both in initiation and in implementation.

Questionnaires were developed by the project team and then pre-tested under actual field conditions. The pre-testing revealed several problems in questionnaire design, which led to revised question formats. Questionnaires were in English, with the field officers (and, when possible, accompanying local experts) translating them in the field. Most commonly, questions and answers were done in Nagamese, the non-written 'lingua franca' used in Nagaland among people from different ethnic backgrounds who do not speak English. Field officers were trained in questionnaire methods and translations agreed upon before going into the field. Surveys were successfully conducted and the three questionnaires collected in all 28 villages. Interpretation of the survey results below is conditioned by five years of interactive work with villagers in Nagaland, using a variety of participatory and informal approaches.

Impact on the community

Elders in the selected villages were informed about the upcoming visit and asked to assemble information about tree planting in their villages, including a list of households that had begun planting trees in their swidden plots after establishing their two test plots. Table 3 presents summary statistical data for several key variables in the questionnaire: village size (number of households), estimated number of households that have begun planting trees in swidden fields (that is, replications), estimated area planted in the replicate plots, and number of trees planted in the replications.

Village size averaged 266 households with the size in the sample varying between 33 and 1200 households. On average, 98 of the households (37 per cent) have planted trees in *jhum* fields since the project was implemented in 1995. A total of 1917 ha of trees were planted in replications in the 28 villages; the average area per village was 69 ha.

Table 3: Summary statistical data for sample (number of observations: 28)				
	Households in villages sampled (no.)	Replications in villages sampled (no.)	Area planted in replications (ha)	Trees planted in replications (thousands)
Total	7441	2731	1917	4180
Mean	266	98	69	149
Standard deviation	258	87	118	160
Maximum	1200	350	565	800
Minimum	33	5	6	10

Extrapolating from the sample to the entire population of villages with test plots at the time of the survey suggests that roughly 59,000 ha of tree plantation (69 ha per village, 854 villages) has occurred subsequent to the project, confirming the casual field observation of widespread tree planting in recent years. However, the standard deviation of mean area replicated per village was very high (118 ha) because of two large outliers – one village reporting 372 ha and another 565 ha. Both outliers were located at very low altitudes with exceptionally good access to forest processing facilities and rail transport. Excluding those two outliers provides a mean of 38 ha per village (standard deviation = 31 ha). Thus, a more conservative estimate of the total area of replication is 32,000 ha (38 ha per village, 854 villages).

Village elders were also asked to estimate the number of trees planted in replicate plots. The project management normally helped to arrange for seedlings for test plots, but villagers usually arranged for their own planting material in replicate plots. By early 1999, when the survey was conducted, seedlings from the limited government programmes available at that time had been made available to a limited number of villages in some, but not all, regions of Nagaland. Many villagers, especially more prosperous ones, reported purchasing seedlings from commercial nurseries. In other cases, local nurseries using village resources were established to provide seedlings, or farmers used planting material (saplings and seeds) collected in the wild. In total, the estimated number of trees planted in replication plots in the sampled villages was nearly 4.2 million trees, averaging 149,000 trees per village. If these self-reported estimates are accurate, extrapolation to the 854 villages included in NEPED indicates that about 130 million trees could have been planted in scaling-up activities since the NEPED

Project began. Excluding the two large replicate plots, the other 26 villages in the sample planted an average of 119,000 trees in replicate plots; extrapolated to the 854 villages, this would indicate that 100 million trees were planted in scaling-up activities.

One concern is that enthusiasm for tree planting might actually increase deforestation because land committed to trees will not be available for food crops after the normal fallow period unless farmers are willing to harvest trees when they are well below optimal market size. Although the market for timber in India appears buoyant and capable of absorbing the trees from Nagaland, serious market assessments were not done, a fault being corrected in the second phase of the project. In each village, villagers set land aside as a forest reserve. This reserve land is not cultivated but is instead used for collecting firewood, hunting, and foraging for other non-timber forest products. These reserves are usually located on steep mountainsides that are inferior for cultivation and are close to the villages. Mature trees may be selectively harvested for timber, but villagers traditionally manage the forest reserve carefully. The remainder of forest land owned by the villagers is normally exploited for crop cultivation, with long swidden cycles when land is abundant and short cycles when it is relatively scarce. Excepted is steeply sloped land at high altitudes (generally above 2000–2500 m). Therefore, to meet village food needs farmers must

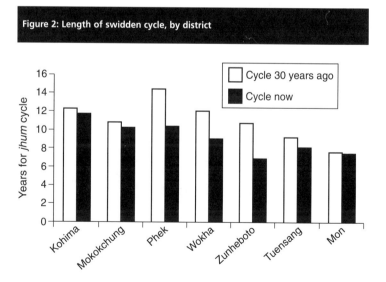

Figure 2: Length of swidden cycle, by district

either intensify the use of available land by shortening swidden cycles or exploit their forest reserves. Figure 2 shows the length of swidden cycles in each district, now and 30 years ago, as reported by village elders. In all cases, cycles have shortened, but particularly so in Phek, Wokha, and Zunheboto Districts. In four cases (Wokha, Zunheboto, Tuensang, and Mon) swidden cycles are now below the benchmark of 10 years needed for the system to be sustainable. In those districts, entire mountainsides are often cleared for swidden cultivation. In Phek and Kohima Districts villagers depend upon irrigated rice production in terraces for most of their food supply and have extensive forest reserves.

When questioned, only three villages reported encroaching on their forest reserves. One village in Kohima District (a village reporting substantial scaling up, 370 ha) reported using over 150 ha, while two other villages in other districts reported 6 and 2 ha of new deforestation, respectively. Thus, the immediate impact on deforestation appears to be nil in most cases and significant in only one village. Extensive field observation by the project field team over five years suggests that the survey results are broadly accurate – there is occasional and generally limited deforestation with a few cases of fairly extensive forest loss. However, the longer-term impacts might be more substantial. In 23 cases, village elders reported that they did not anticipate that tree planting would adversely affect village food supply, and in all 28 cases they reported that their intention was to expand the extent of tree planting in their *jhum* fields.

Most individual farmers who were questioned were also unconcerned about an adverse effect of agroforestry on food security. One reason for this lack of concern might be because of the way in which planting timber trees was generally adopted in Nagaland – as an additional crop in a highly diversified subsistence food mix. Village elders in about 40 per cent of villages indicated that they plan to use village development board funds for planting trees in the future. Their assumption that planting more trees will little affect subsistence food production in the village may be correct, given that only one *jhum* cycle will need to be shortened to compensate for the increased land devoted to planting trees before they realise the anticipated cash income from timber sales. However, if village elders have underestimated the impact on food production that shortened cycles may have, ultimately the village may be forced to increase the use of forest reserves simply to produce enough food.

In our frequent field visits during the course of the project, we observed that local village governments increasingly used village resolutions to plant trees as a way to stimulate villagers to move from swidden food production to combined forestry-and-food systems. Village resolutions are collective decisions taken by village councils. They are usually a powerful force in shaping land use. In the Naga village governance system, village resolutions are commonly a means of initiating new practices or principles of livelihood. The village councils use them as moral persuasion to encourage village members to follow new practices that deviate from well-established traditions. New concepts and practices are difficult to implement in the Naga form of social cohesion without the blessing of village elders, and the village resolution provides legitimacy to newly evolving ideas. We observed that the passing of a village resolution was frequently the formal signal from elders to villagers and the needed stimulus to move from experimentation to widespread adoption.

Twelve of the villages in the sample have passed resolutions to plant trees. Six of these pre-date the implementation of NEPED, with one passed as early as 1985. This is not surprising because the initial impetus for NEPED arose locally from several villages and was broadly promoted by several community leaders. Dividing the sample data into two groups, one comprising villages that have passed a resolution and one group that has not, suggests that village resolutions seem to be effective agents for motivating the integration of trees in swidden fields. In villages with resolutions, 47 per cent of households have adopted the agroforestry system compared with 31 per cent in villages that have not. The average area devoted to agroforestry was 93 ha in the former, compared with 50 ha in the latter. Interestingly, it was the smaller villages that were most successful in passing village resolutions and in stimulating agroforestry. This suggests that size and homogeneity might be factors in determining a group's ability to take action (Mueller 1989). The average size of villages passing a resolution was 215 households (standard deviation of 99), as compared with 304 households (standard deviation of 324) in villages that have not passed a resolution.

Farmer practices

When conducting the survey, a 30 x 30 m square was staked out in the centre of each plot, in which trees were counted and recorded by species. There was little difference between test and replicate plots.

Although there is considerable overall diversity in species planted across Nagaland, farmers placed heavy reliance on teak and gomari (*Gmelina arborea*), which are in strong demand in regional timber markets. More than half of the plots contained gomari, about one-third contained alder, and about one-quarter contained teak.

The average count was 269 trees in the 30 x 30 m squares of the test plots and 213 trees in the 30 x 30 m squares of the replicate plots. Since plot size averaged about 3 ha, about 9000 trees would have survived in each test plot and 7100 in each replicate plot (assuming the same density of trees throughout the plots). This is substantially more than the average 5900 seedlings reported to have been planted in test plots and slightly higher than the 6400 reported planted in replicate plots. Farmers who had test plots were encouraged and trained to weed selectively and allow natural re-growth of valuable species. The data suggest that farmers with test plots appear to have done this, but the technique was not widely disseminated to other farmers.

There was a major difference, however, in the percentage of total trees represented by each species in the 30 x 30 m squares. Replicate plots contained a higher percentage of gomari (*G. arborea*), teak (*Tectona grandis*), hollock (*Terminalia myriocarpa*), and hill toona (*Cedrela serrata*) trees than did test plots; these species generally have a good cash market. These four species accounted for 58 per cent of all trees in replicate plots and only 35 per cent of those in test plots. Test plots contained a greater diversity of species than did replicate plots. About 35 per cent of trees in test plots were not in the top nine listed species, while only 20 per cent of trees on replicate plots were not in the top nine.

Farmers in the survey were asked a series of questions related to their agroforestry practices and the reasons for those practices. Although farmers adopting agroforestry in replications did not have the same level of advisory services as were available to those who had test plots, for the most part they followed similar practices. Farmers were asked about the number of species of trees and seedlings that were planted in their plots, and reported that test plots contained an average of five species while other farmers reported an average of four species.

Only half of the villagers who were awarded a test plot had ever planted trees in a *jhum field* before; 68 per cent subsequently did so in another *jhum* field. Among those who did not receive a test plot, twice as many (86 per cent compared with 43 per cent) planted trees in a *jhum* field after the NEPED test plot was established in their village as had planted trees before. Virtually all villagers indicated that they were likely to plant trees in *jhum* fields in the future.

However, scaling up of the recommended trench land shaping in *jhum* fields was not extensive. Only a limited number of farmers (18 per cent of test plot owners and seven per cent of replicate plot owners) reported that they had made contour trenches in fields after the NEPED test plots were established in their villages. About one-third of the farmers indicated they would possibly use this style of trench land shaping in the future. Most farmers (about 90 per cent) had used traditional erosion control methods before NEPED being established in their villages and reported an intention to continue using these methods. Given that a substantial share of the programme benefits that farmers received was for the land-shaping component, the very limited extent of scaling up suggests that efforts should be concentrated on the more accepted tree-planting component.

Farmers were asked what factors they considered important when making decisions about planting trees in their *jhum* fields. There were no significant differences in the answers given by farmers of test and replicate plots. Expected benefits from tree planting emphasised economic benefits from:

- timber for home use;
- timber for cash sales;
- improved firewood supplies;
- money for children's education;
- security for old age.

However, more than half of the farmers reported that they expected the trees planted to improve soil fertility, provide erosion control, and secure village water supplies. These environmental benefits were given less importance than the direct cash and in-kind consumption benefits from trees.

Lessons

The NEPED project has attempted to stimulate villagers to adopt simple agroforestry systems that integrate timber trees into food crops grown in swidden agriculture. The project concept was originally developed in Nagaland by government officers who were distressed at the results of many unsuccessful attempts to wean villagers from swidden agriculture. People are more inclined to participate in social forestry projects when they perceive some immediate benefits for themselves (Muthayya and Loganathan 1992). When the basic swidden system is augmented instead of attempts being made to radically change it, villagers find it easier to perceive benefits and are more likely to respond, leading to a stronger scaling-up effect.

The approach to project implementation was also radically different from that of previous programmes in Nagaland and in many other regions, which are often centralised, bureaucratic, and top-down in nature. Instead, reliance was placed on local testing and dissemination. Rather than having extension agents arrive in remote villages with a solution in hand, responsibility for selecting, testing, and disseminating agro-forestry systems occurred within the villages. Project officers provided basic training, suggestions, encouragement, and (perhaps most importantly) cash resources to establish test plots, but they allowed villagers broad latitude in selecting the systems they thought most suited to local needs.

The project has been very successful in stimulating replication, as measured by the amount of tree planting that has been integrated into swidden fields in the sample villages. Clear evidence of extensive tree planting can be seen in field visits, the result showing up in the survey data. A conservative estimate is that the 5400 ha of test plots used for local-based testing across 854 villages have been replicated by at least 32,000 ha of tree plantation. This implies a scaling-up rate of at least 6:1. Although replication rates vary, the response has been positive in all districts of Nagaland. Survey respondents were asked to rank the income of project participants and replicators as low, average, or high, relative to typical income in the village. All test plot operators were ranked as middle-income families, while replicators were reported as 89 per cent middle income, 4 per cent high income, and 7 per cent low income.

The actual form of the test and replication plots that were established often differ from the basic concept recommended by project officers. For example, the land-shaping technology has rarely been adopted in replications, most villagers preferring to use indigenous methods of erosion control, such as the water-flow barriers made from available material, when required.

Within villages, replication plots often differ from test plots. In other words, local solutions and adaptations seem to be the norm. Over time, the selection of species planted in test plots narrowed as farmers and project officers gained experience with local species. Limited evidence of encroachment into forest reserves occurred, but generally it was minimal. However, given the plans of all villages in the sample to increase tree plantation and the general shortening of swidden cycles, adoption of agroforestry in Nagaland might still result in additional deforestation and loss of biodiversity down the road, albeit in the relatively benign form of social forestry.

The participatory and highly flexible approach to agroforestry technology created a wide diversity of experiences and outcomes, particularly because the farmer-led programme was implemented on such a broad scale. This diversity – which permitted farmers to first 'pick and choose' in the testing and then to 'watch, learn, and adapt' – was fundamental to scaling up. However, scientific verification is more complex and difficult when implemented on a very large scale, as in NEPED, in a region like Nagaland where transportation and communication are both exceedingly difficult.

This study confirms many of the findings found elsewhere in social forestry (see, for example, Dove 1992, 1995):

- Many foresters believe that small-scale farmers would oppose having trees on their farms because of the long growth period; in reality, some farmers had already experimented with agroforestry and quickly started planting more trees.

- There is an assumption that farmers would be interested only in planting large blocks of market-oriented exotic varieties; in reality, although a few highly valued species are prominent, many farmers also planted multipurpose native trees, such as alder, in substantial quantities. However, although this effect was more pronounced in project-supported plots, farmers who scaled up also tended to plant diversified plots.

- Although farmers are highly motivated by cash-market potential, many also recognise environmental benefits and plant trees to meet household needs for fuel and timber.

- The traditional system of private (individual, family, or clan) property rights in Nagaland was highly likely to be a critical factor in the NEPED success of encouraging scaling up. Farmers who plant trees in Nagaland are assured that they will reap the benefits when trees are ready for market.

- Widespread and regionally large investment in timber trees requires a large and growing market such as that in India. Nagaland is well situated to tap this market.

- The limited success in involving women in agroforestry indicates the impact of property rights. In the limited number of cases where women had clearly defined property rights to land and trees, test plots were properly established and maintained. However, most women's test plots involved temporary and unclear rights, with poor field results. Virtually none of the women's plots was scaled up.

- Initially, the subsidy was considered necessary by the project implementation team as a way to encourage tree planting because this has been the primary tool the Nagaland government had used, and farmers had become accustomed to being paid to try new systems. However, as the project progressed, the speed and magnitude of scaling up provided indications that the need for subsidies had possibly been overestimated.

- Continual monitoring and evaluation (including informal interview techniques with repeated visits) played a critical role in assessing the extent and nature of scaling up, identifying farmer-initiated adaptations, and initiating new project activities to overcome constraints experienced by farmers as the project progressed.

References

Arnold, J.E.M. (1998) 'Managing forests as common property', FAO Forestry Paper 136, Rome: FAO

Baker, J.M. (1998) 'The effect of community structure on social forestry outcomes: insights from Chota Nagpur, India', *Mountain Research and Development* 18(1): 51–62

Beckly, Thomas M. (n.d.) 'Moving Toward Consensus-based Forest Management: A Comparison of Industrial, Co-managed, Community and Small Private Forests in Canada', unpublished report, Edmonton: Northern Forestry Centre

Chatterjee, N. (1995) 'Social forestry in environmentally degraded regions of India: case study of the Mayurakshi Basin', *Environmental Conservation* 22(1): 20–30

Dove, M.R. (1992) 'Foresters' beliefs about farmers: a priority for social science research in social forestry', *Agroforestry Systems* 17(1): 13–41

Dove, M.R. (1995) 'The theory of social forestry intervention: the state of the art in Asia', *Agroforestry Systems* 30(3): 315–40

Egneus, H. (1990) 'Afforestation programmes in India to alleviate poverty through wood production', *Svensk Geografisk Arsbok* 66: 46–59

Faminow, M.D. (1998) *Cattle, Deforestation and Development in the Amazon: An Economic, Agronomic and Environmental Perspective*, Wallingford: CAB International

Higgins, Charlene (1998) 'The role of traditional ecological knowledge in managing for diversity', *Forestry Chronicle* 74(3): 323–6

Jodha, N.S. (2000) 'Poverty Alleviation and Sustainable Development in Mountain Areas: Role of Highland–Lowland Links in the Context of Rapid Globalisation', paper presented at International Conference on Growth, Poverty and Sustainable Resource Management in the Mountain Areas of South Asia, International Centre for Integrated Mountain Development, Kathmandu, 31 January–4 February

Keitzar, S. (1998) 'Farmer Knowledge of Shifting Cultivation in Nagaland', draft report to International Development Research Centre, Mokokchung, Nagaland: State Agriculture Research Station

Mallik, Azim U. and Hafizur Rahman (1994) 'Community forestry in developed and developing countries: a comparative study', *Forestry Chronicle* 70(6): 731–5

Mueller, D.C. (1989) *Public Choice II*, Cambridge: CUP

Muthayya, B.C. and M. Loganathan (1992) 'Community participation in social forestry: a dialogical assessment', *Journal of Rural Development (Hyderabad)* 11(6): 729–46

NEPED and IIRR (1999) *Building upon Traditional Agriculture in Nagaland, India*, Nagaland: Nagaland Environmental Protection and Economic Development, and Silang, Philippines: International Institute of Rural Reconstruction

Ramakrishnan, P.S. (1993) *Shifting Agriculture and Sustainable Development: An Interdisciplinary Study from North-Eastern India*, Delhi: UNESCO and OUP

Ramakrishnan, P.S. (1996) 'Biodiversity research agenda for the Asia-Pacific region', in I.M. Turner *et al.* (eds) *Biodiversity and the Dynamics of Ecosystems*, Singapore: DWPA

Sánchez, P.A. (1976) *Properties and Management of Soils in the Tropics*, New York: John Wiley

Tamale, E., N. Jones, and I. Pswarayi-Riddihough (1995) *Technologies Related to Participatory Forestry in Tropical and Subtropical Countries*, World Bank Technical Paper No. 299, Washington DC: World Bank

Scaling up the use of fodder shrubs in central Kenya

Charles Wambugu, Steven Franzel, Paul Tuwei, and George Karanja

The low quality and quantity of feed resources is the greatest constraint to improving the productivity of livestock in sub-Saharan Africa (Winrock International 1992). Milk demand and production are concentrated around towns and cities where marketing costs are relatively low. Furthermore, farm sizes are also smaller in these peri-urban areas, which exacerbates feed constraints. Fast-growing leguminous trees or shrubs (the terms 'tree' and 'shrub' are used synonymously in this paper) have the potential to alleviate farmers' feed problems. Leguminous trees and shrubs have root nodules that can often fix nitrogen from the atmosphere, making it available to plants. Fodder from these shrubs is rich in protein and, unlike grasses, the shrub leaves maintain their levels of protein even during the dry season. Moreover, farmers can use the shrubs for many other purposes – for hedges along boundaries and around the homestead, for prevention of soil erosion along contours, and for fuelwood.

Since the early 1990s, the National Agroforestry Research Project (NAFRP), based at the Kenya Agricultural Research Institute (KARI) Regional Research Centre, Embu, has been actively testing *Calliandra calothyrsus* fodder shrubs around Embu. The project is jointly managed by the Kenya Forestry Research Institute (KEFRI), and the International Centre for Research in Agroforestry (ICRAF). By 1997, about 1000 farmers in surrounding on-farm trial sites had planted *Calliandra*, but the project lacked the staff and resources required to extend the planting to other areas of the Kenyan highlands. This paper reviews the efforts of a project financed by the Systemwide Livestock Programme (SLP) of the Consultative Group on International Agricultural Research (CGIAR) involving ICRAF, KARI, and the International Livestock Research Institute (ILRI) in facilitating the dissemination of fodder shrubs in the highlands of central Kenya.

Description of study area

The coffee-based land-use system of central Kenya, ranging in altitude from 1300–1800 m, is located on the slopes of Mount Kenya. Rainfall occurs in two seasons, March–June and October–December, and averages 1200–1500 mm annually. Soils, primarily nitosols, are deep and of moderate to high fertility. Population density is high, ranging from 450 to 700 persons/km². In the Embu area, farm size averages 1–2 ha. Most farmers have title to their land, and thus their tenure is relatively secure. The main crops are coffee, produced for cash, and maize and bean, produced for consumption. Most farmers also grow Napier grass (*Pennisetum purpureum*) for feeding their dairy cows and crop their fields continuously because of the shortage of land. About 80 per cent have improved dairy cows, 1.7 cows per family, kept in zero- or minimum-grazing systems. Milk is produced both for home consumption and sale. Forty per cent of the farmers also have goats, averaging 3.2 per family (Minae and Nyamai 1988; Murithi 1998).

The main feed source for dairy cows is Napier grass, supplemented during the dry season with crop residues, such as maize and bean stover, banana leaves and pseudostems, and indigenous fodder shrubs. Forty-five per cent of the farmers buy commercial dairy meal (nominally 16 per cent crude protein) to supplement their cows' diet (Murithi 1998). Farmers complain that the price ratio between dairy meal and milk is not favourable, that they lack cash for buying dairy meal, that its nutritive value is suspect and highly variable, and that it is difficult for them to transport dairy meal from the market to the homestead (Franzel *et al.* 1999).

Research on fodder shrubs

Research on *Calliandra* began in Kenya in the 1980s, by ILRI and KARI. In the early 1990s, NAFRP began conducting on-farm trials with farmers to find out which niches they preferred for planting the shrubs. Farmers did not plant shrubs in pure-stand plots, because of the limited size of their farms, but they found ample space for hundreds of shrubs in hedges around the homestead, external and internal farm boundaries, along contour bunds, or intercropped between rows of Napier grass. Researchers and farmers found that when shrubs are cut at a height of 0.6–1.0 m, biomass yield is substantial and there is little competition with adjacent crops. A farmer managing the shrubs in this way would need about 500 to feed a cow throughout the year at a rate of 2 kg dry matter (equivalent to 6 kg fresh weight) per day, providing about 0.6 kg crude protein (Paterson *et al.* 1996b). The shrubs are first pruned for fodder 9–12 months after planting, and pruning continues at the rate of four or five times per year (Roothaert *et al.* 1998).

Calliandra seedlings are raised in nurseries and transplanted following the onset of the rains. Experiments on seedling production have confirmed that plants may be grown in raised seedbeds rather than by the more expensive, laborious method of planting in polythene bags (O'Neill *et al.* 1997). On-farm feeding trials have also confirmed the effectiveness of *Calliandra* both as a supplement to the cow's diet and as a substitute for dairy meal. The trials found that 1 kg of dry *Calliandra* had about the same amount of digestible protein as 1 kg of dairy meal; both increased milk production by roughly 0.75 kg under farm conditions, but the response varied depending on such factors as the cow's health and the quality of its basal diet (O'Neill *et al.* 1995; Paterson *et al.* 1996a). Researchers are also conducting studies on other shrub species, exotic and indigenous, to help farmers further diversify their feed sources. These species include *Leucaena trichandra*, *Morus alba* (mulberry), and *Sapium ellipticum*.

Scaling up fodder shrub use: achievements and impact

The NAFRP helped farmer groups in the Embu area to set up 14 *Calliandra* nurseries in 1997, 26 in 1998, and 12 in 1999. But extension work was outside the project mandate; therefore, a new project financed by SLP recruited a dissemination facilitator in 1999 to scale up the use of fodder shrubs in central Kenya (ILRI 2000). The scaling-up task was not exclusively to transfer knowledge of fodder shrub technologies and seed to new areas but, equally important and more time-consuming, it was intended to:

- build partnerships with a range of stakeholders in new areas;
- assess whether feed shortage was perceived to be a problem among farmers, gauge their interest in planting fodder shrubs, and determine whether the shrubs were appropriate in their environment;
- assist farmer groups and communities to be effective in mobilising local and external resources for establishing *Calliandra* nurseries; and

- ensure the effective participation of farmer groups and stakeholders in testing, disseminating, monitoring, and evaluating the practice. These tasks were considered vital to ensuring that scaling up would be sustainable once the project was implemented.

Initially, project staff reviewed secondary information and results of farmer surveys to assess appropriate areas for fodder shrubs. Potential collaborating organisations across seven districts (a district comprises roughly 2000–4000 km^2 and 200,000–500,000 people) were identified, including government departments, NGOs, churches, and community-based organisations (CBOs). Fortunately, most were already using participatory research and development methods and confirmed that many of the farmers they worked with had critical problems feeding their dairy cows and were interested in planting fodder shrubs. Farmers in some areas, such as those focusing on irrigated vegetable production, were not interested in planting fodder trees.

Project activities extended across seven districts but were focused in clusters within each district to reduce costs and to facilitate monitoring and the exchange of information among groups. Meetings were held with farmers to discuss the problems they had in feeding their cows and to explain to them the costs, benefits, and risks of planting fodder shrubs. Farmer visits were arranged to see farmers in the Embu area who had already had several years of experience in growing and feeding *Calliandra* to their dairy cows. Most of the farmer groups paid for their own transport and subsistence costs on these visits. Seeing and discussing *Calliandra* with experienced farmers was an effective means to promote *Calliandra* planting and to provide a forum for farmers to learn about its growth, management, and use. The tours involved 420 farmers from 25 groups and 20 extension staff.

For areas where farmers were interested in fodder trees, project staff and partners discussed the terms of collaboration and each party's role was made explicit: SLP staff would initially provide the training and seed but after two to three years the partner organisation would take over these functions. Joint workplans were then developed, which clearly indicated a schedule of training events and follow-up activities.

Needs assessments were undertaken to determine farmers' knowledge and skills and to ensure that training would build on farmers' indigenous knowledge. Once farmers were trained to establish nurseries, they, in turn, trained their neighbours. Farmers in the clusters were also trained in seed production so they could provide seed to neighbouring farmers and to extension staff for distribution in other areas.

In 1999–2000, the project assisted staff of the following organisations to help farmers establish nurseries: the provincial administration in two provinces, three departments of the Ministry of Agriculture and Rural Development, one international NGO, four local NGOs, one extension service of a private company, two church extension services, ten CBOs, and 150 farmer groups. Farmer groups ranged in size from four to 50 members, and averaged about 17. Most of the groups were already in existence before the project, promoting such activities as dairy goats, handicrafts, domestic water tanks, soil conservation, organic farming, and shrub nurseries. Most (76 per cent) of the groups included both men and women; 15 per cent were women's groups, and 9 per cent were men's groups (see Table 1). Women accounted for 60 per cent of all group members. Most groups had more than one nursery. Nurseries were located on the farm of a member who had access to water during the dry season, which was essential for successful nurseries. Group members divided the labour amongst themselves and shared the seedlings produced. Ten nurseries were also established in school or church compounds and served as demonstration sites for farmers in the area.

By the end of 2000, the 150 groups had developed 250 nurseries involving over 2600 farmers (see Table 2). On average, farmers each transplanted about 400 *Calliandra* seedlings, of which about 240 (60 per cent) survived. Drought was the main cause of the high mortality.

Table 1: Types of farmers and groups establishing fodder shrub nurseries in the central highlands of Kenya

Type of farmer	No.	%	Type of group	No.	%
Female	1560	60	Mixed groups	115	76
Male	1040	40	Women's groups	22	15
Total	2600	100	Men's groups	13	9
			Total	150	100

Table 2: Expansion in numbers of farmer groups planting fodder shrub nurseries in the central highlands of Kenya

Season and year	No. of districts	No. of farmer groups	No. of nurseries	No. of farmers
1999 long rains	2	12	12	220
1999 short rains	6	117	180	2037
2000 long rains	7	150	250	2600

Rainfall was lower than normal during three consecutive seasons: the short rains of 1999, and both the long and the short rains of 2000.

Selected group members were trained in how to produce and distribute seeds. *Calliandra* begins producing seed in its second year but unfortunately the shrubs produce relatively little seed, and collecting it is laborious. Some farmers and private nurseries have begun selling *Calliandra* seed and seedlings, and the numbers doing so are likely to increase as production and demand for the shrubs increases.

Dependence on a single fodder shrub species is risky. Diversification reduces the risk of pest and disease attack, improves feed quality, and increases biodiversity. Therefore, the project has started disseminating other fodder shrub species; farmers in 80 groups have planted *L. trichandra*, 70 groups have planted *M. alba*, and 13 have planted a herbaceous legume, *Desmodium intortum*.

Impact assessment of this initiative has not yet been carried out, but an economic analysis was conducted of farmers' *Calliandra* fodder banks in the farmer-managed on-farm trials around Embu (Franzel *et al.* 1999). The analysis indicated that beginning in the second year after planting, a farmer with 500 shrubs would earn an extra US$130 per year, either through increased milk production or through reduced purchase of dairy meal. If 50 per cent of Kenya's estimated 625,000 smallholder farmers owning dairy cows each planted 500 fodder shrubs, the net benefits per year would reach US$81 million (Franzel *et al.* 1999).

The impact can also be important for farmers with dairy goats, an enterprise that is particularly well suited for farmers lacking the resources to buy and feed a dairy cow. Dairy goat production is growing rapidly in Kenya, and about 1300 dairy goat farmers in the highlands of central Kenya have planted *Calliandra*. Their feedback has confirmed the results of experiments at KARI-Embu, which found that *Calliandra* is an excellent feed for dairy goats (ILRI 2000).

Monitoring, farmer innovation, and feedback

Informal monitoring takes place in which farmers and extension staff provide feedback to project staff and researchers on their progress and problems. In one case, feedback on a farmer innovation has resulted in a change in recommendations made by extension services. Farmers in Kandara Division, Maragua District, conducted experiments on soaking *Calliandra* seeds before planting and found that seeds soaked for 48–60 hours had higher germination rates than those soaked for the recommended 24 hours. Researchers at KARI-Embu confirmed the farmers' findings and extension staff now recommend the longer soaking time.

Farmers' problems with pests and their innovations in controlling them have also led to the design of new on-farm trials. For example, in 2001, researchers and farmers are comparing the effectiveness of using netting and local measures (spraying solutions made from tobacco, marigold, neem, hot pepper, or *Tephrosia vogelii*) to control crickets, hoppers, and aphids damaging seedlings in nurseries. These findings demonstrate the importance of monitoring farmers' innovations and feeding them back to research and extension.

Formal questionnaire surveys began in 2000 to assess farmers' experiences with *Calliandra*, problems encountered, and factors explaining adoption and successful group and nursery performance. The surveys are conducted with funding from the CGIAR Systemwide Programme on Collective Action and Property Rights. Results are not yet available, but because researchers from KARI-Embu are involved in conducting the surveys they are expecting considerable feedback from the field.

Problems encountered

Severe drought and poor distribution of rainfall increased the mortality of seedlings in the nurseries and shrubs in the field. Unlike in many areas of Africa, severe drought during the long rains season is extremely rare in central Kenya. Nevertheless, there is a high demand in 2001 for seed for nurseries, and farmers are being urged to locate nurseries near permanent sources of water. Infestation by crickets, hoppers, and aphids, as mentioned above, has also led to a significant loss of seedlings. These pests are particularly damaging during dry periods. The high turnover among staff of the Ministry of Agriculture, poor morale, and lack of resources such as transportation have also constrained success. The SLP project occasionally assists ministry staff with transportation and subsistence allowances, which greatly increases staff motivation.

Factors contributing to success

Several factors have contributed to the achievements thus far:

- The demand among farmers for fodder shrubs was huge, mainly because the shrubs save cash, farmers' scarcest resource, and require only small amounts of land and labour.

- The project area is noted for the dynamism of its farmers, and access to markets is fairly high, both of which enhance the adoption of new practices.

- Because the project works through partner organisations instead of directly with farmers it is able to build on local organisational skills and knowledge and reach far more farmers than would otherwise be possible.

- Dissemination through farmer groups instead of individual farmers economises on scarce training skills and transport facilities. In addition, working with groups ensures greater farmer-to-farmer dissemination and exchange of information.

- The strong partnership between researchers, extensionists, and farmers in the project facilitates the flow of information among the three.

Remaining challenges

Nevertheless, several critical challenges remain:

- While the project has successfully expanded the use of fodder shrubs across seven districts, it is still reaching only a small percentage of dairy farmers in these districts and less than 1 per cent of Kenya's smallholder dairy farmers. Further scaling up is required, focusing on institutions working in areas of the country where smallholder dairy farmers predominate. ICRAF, the Oxford Forestry Institute in the UK, and other partners are planning a project that will help the Ministry of Agriculture and Rural Development, NGOs, and farmer organisations throughout East Africa to assist farmers to plant *Calliandra* for fodder.

- Commercial seed production and distribution are slowly emerging in project areas, but it is not clear if seed production will continue to grow and meet local demand. Greater emphasis is needed on promoting community-based seed production and distribution through a range of partners: farmer groups, individual seed producers, and private nurseries. The SLP project is beginning work in this area.

- Greater diversification of fodder shrubs is needed to reduce the risk of pest and disease attacks, improve feed quality, and increase biodiversity. KARI-Embu has a strong programme for evaluating fodder trees and is increasing its emphasis on indigenous species.

- A consortium of partners needs to be established for promoting fodder shrubs. While the project is currently the hub of the informal network, providing seed and training, other organisations need to take over these functions in future years. Setting up periodic meetings of partners, including farmers, can help promote the exchange of skills, seed, and information, enhancing the spread of fodder shrubs and increasing household income from dairy. The first such meeting of the consortium is scheduled for 2001.

- Extension materials need to be developed to promote *Calliandra*. Videos and simplified brochures for farmers, such as that by Wambugu (2001), are among the tools that will be most useful.

Finally, experience confirms that successful scaling up of a new practice requires much more than transferring seed and knowledge about it. Rather, facilitators need to build partnerships with and among a range of stakeholders, ensure farmers' interest in the practice and its appropriateness to their conditions, assist farmer groups and communities to mobilise local and external resources effectively, and ensure the effective participation of farmer groups and stakeholders in the processes of testing, dissemination, and monitoring and evaluation.

Acknowledgements

The authors are grateful to Peter Cooper and Chin Ong for reviewing earlier versions of this paper.

References

Franzel, S., H. Arimi, F. Murithi, and J. Karanja (1999) Calliandra calothyrsus: *Assessing the Early Stages of Adoption of a Fodder Tree in the Highlands of Central Kenya*, AFRENA Report No.127, Agroforestry Research Network for Africa, Nairobi: International Centre for Research in Agroforestry

International Livestock Research Institute (ILRI) (2000) *CGIAR Systemwide Livestock Programme: Biennial Report 1999–2000*, Nairobi: ILRI

Minae, S. and D. Nyamai (1988) *Agroforestry Research Proposal for the Coffee-Based Land-Use System in the Bimodal Highlands, Central and Eastern Provinces, Kenya*, AFRENA Report No. 16, Agroforestry Research Network for Africa, Nairobi: International Centre for Research in Agroforestry

Murithi, F.M. (1998) 'Economic evaluation of the role of livestock in mixed smallholder farms of the central highlands of Kenya', PhD thesis, Department of Agriculture, University of Reading, UK

O'Neill, M. *et al.* (1997) *National Agroforestry Research Project, Kenya Agricultural Research Institute, Regional Research Centre, Embu, Annual Report: March 1995–March 1996*, AFRENA Research Report No. 108, Agroforestry Research Network for Africa, Nairobi: International Centre for Research in Agroforestry

O'Neill, M., F. Murithi, *et al.* (1995) *National Agroforestry Research Project, Kenya Agricultural Research Institute, Regional Research Centre, Embu, Annual Report: March 1994–March 1995*, AFRENA Research Report No. 92, Agroforestry Research Network for Africa, Nairobi: International Centre for Research in Agroforestry

Paterson, R.T., R. Roothaert, and I.W. Kariuki (1996a) 'Fodder trees in agroforestry: the Kenyan approach', in J.O. Mugah (ed.) *People and Institutional Participation in Agroforestry for Sustainable Development: Proceedings of the First Kenya National Agroforestry Conference, 25–29 March 1996*, Muguga, Nairobi: Kenya Forestry Research Institute

Paterson, R.T., R. Roothaert, O.Z. Nyaata, E. Akyeampong, and L. Hove (1996b) 'Experience with *Calliandra calothyrsus* as a feed for livestock in Africa', in D.O. Evans (ed.) *Proceedings of an International Workshop on the Genus Calliandra, 23–27 January 1996*, Bogor, Indonesia: Winrock International

Roothaert, R. *et al.* (1998) *Calliandra for Livestock*, Technical Bulletin No. 1, Embu, Kenya: Regional Research Centre, Embu

Wambugu, C. (2001) Calliandra calothyrsus: *Nursery Establishment and Management, A Pamphlet for Farmers and Extension Staff*, Nairobi: International Centre for Research in Agroforestry

Winrock International (1992) *Assessment of Animal Agriculture in Sub-Saharan Africa*, Little Rock AR: Winrock International Institute for Agricultural Development

The Landcare experience in the Philippines: technical and institutional innovations for conservation farming

Agustin R. Mercado, Jr., Marcelino Patindol, and Dennis P. Garrity

Contour hedgerow systems using nitrogen-fixing trees have been widely promoted as important components of soil conservation in South-East Asia in order to minimise soil erosion, restore soil fertility, and subsequently improve crop productivity. Although positive results have been observed and reported in a number of experimental and demonstration sites, farmers have been slow to adopt the systems. A number of factors are believed to cause this slow adoption: the high amount of labour needed to establish and manage the hedgerows; the poor adaptation of leguminous trees to acid upland soils; the lack of ready sources of planting materials; and the fact that hedgerows may reduce crop yields through their strong above- and below-ground competition with the crop.

ICRAF has been conducting research on contour hedgerow technologies in Claveria, northern Mindanao, Philippines, for the past decade. Intensive examination of many facets of contour hedgerow systems has led to the conclusion that hedgerow systems of leguminous trees consistently increase maize yield by 20–30 per cent, although reasonable yields cannot be maintained without external nutrient supply (particularly of phosphorus [P]) in addition to the tree prunings. However, the yield increase realised does not sufficiently compensate for the extra labour needed to establish and manage the tree hedgerows. Thus, net returns to the practice are usually low. The result is that tree hedgerow systems are usually abandoned after several years of trial.

This does not imply that farmers are not concerned about soil erosion. Erosion was, in fact, one of the top concerns among farmers in our surveys. What it does imply is that any technology, to be adopted, must have minimal cost to farmers, as well as to the public institutions supporting the programme.

Agroforestry or soil conservation technologies must fit within the context of marginal farmers, and be based in their socio-economic and biophysical environments. The socio-economic environment includes land, labour, and capital. The biophysical environment includes soil, climate, and vegetation. Any agroforestry or soil conservation technology must promote plant species that can adapt to the soils of the upland farmers, which are often poor and vary from site to site. Therefore, there is a strong need to develop technology options that consider such complexities.

This paper focuses on two issues: (1) the elements of a low-labour and low-cost system of buffer strips as an approach to conservation farming in the uplands, which may evolve into more complex agro-forestry systems; and (2) institutional innovations based on farmer-led organisations that empower the community and the local government to disseminate conservation farming and agroforestry practices effectively and inexpensively.

The farming systems in the uplands of Claveria, as is typical in many parts of Mindanao, are based predominantly on two crops of maize per year. Farm size averages 3 ha. Tillage is done with animal power. Most farmers are clearly aware of the reasons for declining crop yields and possible strategies to combat soil degradation. Sloping fields in Claveria, which receive 2200 mm of rainfall a year, may lose up to 200 tonnes of soil per hectare. About 59 per cent of the cropping (mostly maize and some vegetable farming) is done on lands of more than 15 per cent slope (Garrity and Agustin 1994; Fujisaka *et al.* 1994). As is typical for the majority of cultivated upland areas in South-East Asia, soils in Claveria are degraded and acidic (pH 4.5–5.2) with low available phosphorus.

Contour hedgerows of pruned leguminous trees, known locally as sloping agricultural land technology (SALT), had been promoted in Claveria since the early 1980s by the Philippine Department of Agriculture as a solution to the problems of unsustainable crop production in the uplands. This farming system aimed to provide effective soil erosion control, organic fertiliser for the companion annual food crops, fodder for the ruminants, and fuelwood, and to restore water quality and quantity in the watershed. In spite of these benefits, adoption by farmers was not widespread. After years of ICRAF's on-farm research, working closely with farmers, we identified the key problems with their use:

- There were high labour requirements in order to establish and maintain the hedgerows.
- Farmers experienced only limited improvement of farm income.
- Unanticipated problems occurred in soil fertility because the hedgerows competed with the annual crops for nutrients particularly in phosphorus-deficient acid upland soils.
- The irregular width of the alleys makes inter-row tillage difficult, because this is done using animal traction.
- There was a reduction in the area available for cultivation because hedgerows were spaced too closely together on moderately to steeply sloping farms, also there was poor species adaptation and a lack of suitable planting materials.
- Farmers often have insecure land tenure.

We were probably very fortunate when we started working in Claveria in 1985 to have had no experiment station upon which we might have conducted our trials on tree legume hedgerows. If we had, we might still be a couple of cycles behind where we are now in our learning experience. Working with farmers on experiments that were super-imposed on contour hedgerows that they had installed themselves made it clear that pruned tree hedgerows were too labour intensive, and productive forage grass hedgerows were too competitive with the associated crops. Neither technology was being adopted. However, we saw that the concept of contour hedgerows was popular. We observed that some farmers experimented with the concept by placing their crop residues in lines on the contour to form 'trash bunds'. These rapidly re-vegetated with native grasses and weeds and soon formed stable hedgerows with natural front-facing terraces. Other farmers tried laying out contour lines but did not plant anything in them. These lines evolved into natural vegetative strips (NVS), which we later observed were superb in controlling soil erosion, and required little maintenance (Garrity 1996; Agus 1993).

These latter innovations caught the imagination of many more farmers. By about 1994, over 150 farmers had adopted contour hedgerow systems, while the number of pruned-tree hedgerow fields was by that time decreasing. Meanwhile, the number of farmers with natural vegetative strips continued to increase spontaneously, with adoption spreading from farm to farm. We also observed a broad-based change in tillage systems. When research had first begun in Claveria in 1985, virtually all farmers ploughed up and down the slopes. Contour ploughing was unheard of. By 1995 it was evident that nearly all farmers had converted to the idea of contour ploughing, or were at least attempting to do so.

Evolving components of a successful conservation farming system

Interest in NVS continued to increase. Since it is quite uncommon for large numbers of farmers to adopt an effective soil conservation structure spontaneously, and without public subsidy, we realised that perhaps we were witnessing the kind of low-labour, zero-cash-cost alternative that would be widely applicable. We began to examine each component of the process of establishing and maintaining low-labour hedgerow practices. Establishing NVS requires only a fraction of the labour needed to establish the conventional contour hedgerow of tree legumes. Laying out contour lines, about two person-days per hectare, is all that is required. The total time needed for ploughing is reduced in proportion to the area of unploughed strips. This reduction offsets the labour spent for laying out the contour strips. The amount of labour required to prune or maintain the NVS is proportionate to the spacing of hedgerows. Mercado *et al.* (1997) found that NVS spaced 6 m apart, and dominated by *Chromolaena odorata*, required 15 person-days per cropping per hectare or 30 person-days per year to maintain. This was less than a quarter of the time required for conventional contour hedgerow systems based on tree legumes (ICRAF 1996). Low-statured NVS like *Paspalum* spp. or *Digitaria* spp., require even fewer days (three to ten per cropping season) (Mercado *et al.* 1997; Stark 1999).

Our surveys of farmers who had not yet installed contour hedgerow systems but wanted to do so indicated that their overriding reason for not contouring was that they lacked the technical know-how. We had recently uncovered an extremely simple and practical means of laying out contours without equipment such as an A-frame – namely, the 'cow's back' method (ICRAF 1996). This method involves ploughing across the slope and maintaining the angle of the cow's back on the level. When the animal is heading upslope, its head is higher than its back; when it is off-course downslope, the rear part of the animal is elevated above the front. Stark *et al.* (2000) found that farmers using the cow's back method deviated on average less than 2 per cent from the real contour, compared with either the A-frame method or the hose level method. This deviation is quite acceptable for practical purposes, particularly in light of the fact that most farmers do not bother with A-frames at all, but simply judge the contours visually (which is much less accurate).

Feedback from farmers also elucidated another factor that causes many smallholders to hesitate in installing contour hedgerow systems. Conventional recommendations indicate that hedgerows be separated by a drop of only 1–1.5 m in elevation. On steep slopes, the crop area lost to the strips might thus be 15–20 per cent or more. Crop yields cannot be expected to increase enough to counterbalance this quantity of area lost. Labour also increases in establishing and maintaining many strips in each field. We therefore conducted trials to determine how reducing the density of buffer strips would affect the loss of soil. We found that strips spaced at a vertical drop of 4 m are still effective in reducing soil loss (Mercado *et al.* 1997). Even a single NVS strip placed on the contour half-way down a slope 60 m long reduces soil loss to 40 per cent of that on the open slope. We conclude that farmers could space their strips at much wider intervals than the conventional rule-of-thumb recommendation suggests, even up to 8–12 m apart on such slopes. Erosion control will not be quite as good, but the practice is very much more likely to be adopted. More strips can always be added in between the original ones after the farmer has gained confidence in the effectiveness of the practice.

This wider spacing is also particularly appropriate when the farmer intends to convert the NVS strips into fruit or timber trees, in which there is now great interest in Claveria. To do this, farmers establish contours, then raise their tree seedlings. They introduce the trees during the second or third year after the NVS are established. Tree canopies start to close three to four years later, when the NVS are narrow (<8 m). By this time it is no longer feasible to plant annual crops because the alley is too shady. Some farmers bring in ruminants to graze under the trees.

Farmers with wider alleyways (8–12 m) can still plant annual food crops between the rows of the trees and grow fodder grass between trees along the row. A wider spacing of NVS is useful for farms that want to continue growing food crops while the fruit and timber trees mature. However, farmers with larger farm sizes tend to opt for somewhat closer buffer strip spacing, and cultivate their food crops on other land parcels once the tree canopy shades the annual crops. The fast-growing timber tree systems have a six- to eight-year cycle.

Farmers who establish cash perennial hedgerows such as coffee tend to space hedgerows more closely in order to have more rows of these crops. The cash crops from the buffer strip component often earn more than the maize or other annuals planted in the alleys – NVS can evolve into many forms of agroforestry systems. Farmers in Claveria are planting fodder grasses and legumes, 31 species of timber and fruit trees, and other cash perennials on their NVS fields. The fodder grasses used include *Setaria* spp., *Pennisetum purpureum*, and *Panicum maximum*. The forage legumes include *Flamingia congesta* and *Desmodium rensonii*. Timber species cultivated include *Gmelina arborea*, *Eucalyptus* spp., *Sweitienia* spp., and *Ptericarpus indicus*. The fruit species include mango, rambutan, durian, pineapple, and banana. The wide diversity of species helps the farmers to stabilise their income.

The groundswell of enthusiasm among thousands of Claveria farmers, and the rich store of farmer experiences with a wide range of prospective buffer strip management options, provided a stimulus. Public-sector research and extension institutions needed to consider how they might evolve more effective techniques to diffuse NVS technology rapidly to much larger numbers of interested farmers. The adoption and technology modification process was well documented by IRRI staff (Fujisaka 1989; Cenas and Pandey 1995), but this was not followed by any quantity of extension work.

Extension methods can be basically classified as belonging either to an individual (or household) approach or to a group approach. The former is most effective for activities to be undertaken within the full control of the individual farmer or household (such as establishing contour buffer strips). Working with groups or the community at large is more suitable for matters related to the whole community (such as post-harvest public grazing practices) or for activities that would be undertaken more cheaply by a group (such as tree nurseries). The latter approach is particularly suitable where group work is common. This is practised in the Philippines through the *bayanihan* system, which involves farmer work groups based on voluntary work contribution for a common benefit.

Towards effective technology dissemination: the evolution of an innovative extension strategy

In addition to conducting applied research, ICRAF recently initiated a technology dissemination programme to ensure that derived innovations will reach the user group. ICRAF is helping to strengthen existing government programmes and to help technology dissemination develop into a self-perpetuating farmer initiative. The key institutional innovation in these effects is the Landcare approach: a process that is led by farmers and community groups, with support by the local government and technical backup from ICRAF, from government line agencies such as the Department of Environment and Natural Resources, the Department of Agrarian Reform, and the Municipal Agriculture Office, and from NGOs.

What is Landcare?

Landcare is a method for diffusing agroforestry practices rapidly and inexpensively among upland farmers, based on farmers' innate interest in learning and sharing knowledge about new technologies that enable them to earn more money and to conserve natural resources (Garrity and Mercado 1998). Landcare groups bring together people who are concerned about land degradation problems and interested in working together to do something positive for the long-term health of the land. It evolved as a participatory community-based approach designed to bring about change in complex and diverse situations (Swete-Kelly 1997).

The Landcare model has a threefold emphasis: appropriate technologies, effective local community groups, and partnership with government (Campbell and Siepen 1994). This grassroots approach is generally recognised as a key to success in all community development activities. Groups respond to the issues that they consider locally important, solving problems in their own way. Landcare depends on self-motivated communities responding to community issues, rather than to issues an external agency imposes. Such bottom-up approaches are more likely to bring about permanent and positive change. Landcare groups have government support, and they use networks to ensure that ideas and initiatives are shared and disseminated.

In 1996, ICRAF supported dissemination activities in Claveria as a direct response to the farmers' request for technical assistance in conservation farming. The technical and institutional innovations led to the formation of the Claveria Landcare Association. Today, there are 250 Landcare groups in the municipalities of Claveria, Malitbog, and Lantapan in northern and central Mindanao. Most of these Landcare groups are based in the *sitio* (subvillages) where farmers can interact with each other more frequently. More than 3000 farming families are now involved in these three municipalities alone.

The Landcare groups in Claveria have successfully extended conservation farming based on NVS to an additional 1500 farmers. They have established more than 300 communal and individual nurseries, which produce hundreds of thousands of fruit and timber tree seedlings that are planted on the NVS or along farm boundaries. They have also been able to link to other service providers to get funding for livelihood projects.

Steps involved in the Landcare approach

Based on the evolution of Landcare during the past several years in Claveria, we have identified the major principles and steps in developing this approach (Garrity and Mercado 1998):

1. *Select appropriate sites to bring conservation farming technologies to where they are needed most – on sloping lands where soils are subject to erosion and degradation.* This initial step also involves meeting with key leaders in the local government units (municipal or province), interested farmers, and other stakeholders. Their understanding of the issues that need to be addressed, as well as their willingness to support and complement the programme, are crucial to the success or failure of Landcare at a given site.

2. *Expose key farmers to successful technologies and organisational methods.* This helps to develop strong awareness among prospective core actors – especially innovative farmers and farmer leaders – of the opportunities to address production and resource conservation objectives effectively through the new technologies. The success of the activities can be measured by how much enthusiasm develops within the community to adopt the technologies. Exposure activities include:

   • organising cross-visits to the fields of farmers who have already adopted and adapted the technology successfully into their farming systems;

   • providing training for farmers in the target communities to learn about the practices through seminars in their *barangays* (villages); and

   • providing opportunities for farmers to try out a technology on their land through unsubsidised trials, to convince themselves that it works as expected. These farmers can then become the core of a 'conservation team' to diffuse the technology in the municipality.

The characteristics and roles of farmers, the community, the local government unit, and the technical facilitator in implementing the Landcare approach are listed below:

Farmers:

- are usually resource poor;
- want to improve their livelihood;
- want to employ new farming techniques;
- would like to acquire and share knowledge and experience with other farmers;
- are committed to resource conservation;
- can create work groups for establishing nurseries, conservation farms, etc..

Local government units:

- provide policy support (e.g. institutionalisation of conservation farming and agroforestry, creation of municipal and *barangay* ordinances);
- play a leadership role (e.g. facilitate formation of Landcare groups and activities);
- build capacity (e.g. initiate various training activities);
- facilitate financial support: a Human Ecological Security fund is available from the municipality and from the *barangay*.

Technical facilitators (ICRAF and line agencies):

- develop technology: soil and water conservation, agroforestry, nurseries;
- facilitate formation of Landcare groups and Landcare-related activities;
- provide germplasm;
- initiate information and education campaigns.

3. *Organise local conservation teams.* Once it is clear that there is a critical threshold of local interest in adopting the technologies and a spirit of self-help to share the knowledge within and among the *barangays* of a municipality, the conditions are in place to implement a municipal conservation team. The team is composed of an extension technician from the Department of Agriculture or from the Department of Environment and Natural Resources, an articulate

farmer experienced in the application of the technology, and an outside technical facilitator.

The team initially helps individual farmers implement their desired conservation farming practices. Later, they give seminars and training sessions in the *barangay* if sufficient interest arises. During these events they respond if there is interest in organising more formally to accelerate the spread of agroforestry and conservation practices.

4 *Facilitate a Landcare farmers' organisation.* When the preconditions are in place to form a Landcare farmers' organisation, the facilitator may help the community to develop a more formal structure. A key ingredient of success is identifying and nurturing leadership skills among prospective farmers in vision and organisation. This may involve arranging for special training in leadership and management for the farmer leaders and exposing them to other successful Landcare organisations. Each *barangay* may decide to set up its own Landcare Association chapter and *barangay* conservation team. A *barangay* may organise Landcare Association subchapters in their *sitios* (sub-*barangays*). A *sitio* conservation team usually includes a local farmer-technologist, the *sitio* leaders, and the district *kagawads* (councillors). The *sitio* teams are the frontliners in conservation efforts, providing direct technical assistance, training, and demonstrating to farmer households. They are backed up by the *barangay* and municipal conservation teams.

In the municipality, the Landcare Association is a federation of all of the *barangay* Landcare chapters. The municipal conservation team is part of the support structure, which also includes other organisations that can assist the chapters (for example, the Department of Agriculture, the Department of the Environment and Natural Resources, and NGOs). Figure 1 presents the organisational set-up of the Claveria Landcare Association (CLCA). It is a people's organisation, registered as an association with the Philippine Securities and Exchange Commission (SEC) in 1996.

5. *Attract local government support.* Local government can provide crucial political and sustained financial support to the Landcare Association. The municipality has its own funds earmarked for environmental conservation that can be targeted to Landcare activities. The municipality can be encouraged to develop a formal natural resource management plan – which may help to guide the

allocation of conservation funds. The *barangays* can allocate financial resources from their regular internal revenue allotment through the Human Ecological Security (HES) programme, which represents one-fifth of the total development funds of the *barangay*. These funds can be used to organise the conservation teams and Landcare Association activities in the *barangays* and the *sitios*, and support training activities and honoraria for resource persons if the time required for these activities is more than volunteer time can cover. The municipality can also allocate HES funds to complement the *barangay* budget. For 1998, the Claveria municipal government committed 50,000 pesos (about US$1250) to each *barangay* to support Landcare activities.

External donor agencies can best support Landcare development by allocating resources for leadership and human resources development, communications equipment (such as handheld radio sets), and transportation (e.g. motorcycles) to enable the Landcare leaders to make maximum use of their time.

6. *Monitor and evaluate.* Monitoring is needed to assess progress and make the programme more dynamic and relevant to the needs of the target community. For monitoring purposes, ICRAF has been keeping records of all those who have attended a training session or have been assisted with establishing NVS on their farms, as well as of farmers who have requested assistance. Details on farming and conservation practices, training activities, and follow-up needs are recorded on a diagnostic card, which is updated on regular follow-up visits by ICRAF staff. The leaders of the CLCA chapters or subchapters have been supporting this activity by facilitating the distribution and collection of the diagnostic cards to and from the villages and new CLCA members.

A survey on adoption and dissemination progress is now being conducted, with an emphasis on how farmers modify technologies, and the reasons behind their decision making. A participatory monitoring and evaluation system is being developed that enables Landcare groups to self-evaluate their performance against their objectives. The Landcare facilitators will assist the groups to conduct these exercises, to reflect group accomplishments, and to help groups achieve future goals.

Municipal level

Claveria
Landcare
Association

Actors
- President, Claveria Landcare Association
- Municipal conservation team
- President of all village Landcare chapters
- Municipal mayor
- Chairman, committee on agriculture and environment, Municipal Council
- Municipal agriculture officer
- State College of Agriculture
- ICRAF

Village level

Village
Landcare

Actors
- Village conservation team
- Agriculture technicians
- Chair, Agriculture and Environment Committee, village council
- Village chieftain

Village
Landcare

Subvillage level

Subvillage
(*sitio*)
Landcare

Actors
- Sub-chapter Landcare president
- Subvillage conservation team
- Households
- Agriculture technician
- Chair, Agriculture and Environment Committee
- Subvillage chieftain

Subvillage
(*sitio*)
Landcare

Conservation farming technologies adopted by Landcare members

The specific activities of Landcare Association members will vary according to their needs and interest, as well as their biophysical and socio-economic situation. Some of the many activities that have been or are being developed as focal areas for Landcare Association work include:

- establishing NVS along the contour to reduce soil erosion in the field and on the farm – the initial farmer-generated technology that launched the organisation of Landcare in Claveria;

- planting perennial crops on or just above the NVS to increase the farmers' cash income and enhance soil and water conservation;

- planting trees to increase family income by producing timber, fuelwood, and other tree products in farm forests, boundary planting, or other arrangements;

- planting high-quality fruit trees to provide income and better nutrition for the household while enhancing the environment;

- adopting minimum-tillage or ridge-tillage farming systems; ridge tillage has been successfully adopted with the existing draught-animal cultivation practices and is being further tested on farms;
- establishing nurseries for fruit and timber trees;
- promoting and adopting backyard gardening, thus helping to address the problem of malnutrition, which is widespread among children;
- planting herbal medicines;
- managing solid wastes by segregating the biodegradable wastes and making them into compost;
- setting up local competitions such as composing Landcare songs and slogans to promote awareness and adoption of various resource-conservation measures;
- exchanging labour;
- helping one another in times of sickness, death, and other community problems.

The evolution from simple soil conservation practices to more complex agroforestry systems occurs over time as farmers continually experiment and innovate technologies that are suitable to their conditions. Generally, farmers start by establishing natural vegetative strips. Next, they establish communal or individual nurseries and plant perennials on or above the NVS. Farmers may cultivate annual cereal crops up to the fourth year, particularly if the strips are not too close to each other. When tree canopies shade out the crops and it is no longer profitable to grow annuals, farmers graze livestock beneath the trees. The trees (mostly *Gmelina arborea*) can be harvested 8–12 years after planting, when farmers resume annual cropping and begin the next cycle. This system earns more than the traditional practice of mono-cultural cropping (Magcale-Macandog *et al.* 1997).

Impacts and scaling up

The greatest success of Landcare is in changing the attitude of farmers, policy makers, local government units, and landowners about how to use the land and protect the environment. It is not simply about the total length of NVS laid out, the number of nurseries established, or the number of Landcare members. The Landcare movement is renovating the attitudes and practices of the farmers, policy makers, and local government officials towards using the land to meet their current needs while conserving it for future generations. Now many farmers voluntarily share their time and efforts, while policy makers also urge farmers to adopt conservation farming practices, and support these efforts by allocating local government funds and enacting local ordinances. These are the important success indicators of the Landcare approach that enable local people to conceive, initiate, and implement plans and programmes that will lead to their adopting profitable and resource-conserving technologies. The Landcare approach provides:

- a way for interested farmers to learn, adopt, and share knowledge about new technologies that can earn more money and conserve natural resources;
- a forum in which the community can respond to issues that it sees as important;
- a mechanism for local government to support;
- a network for ensuring that ideas and initiatives are shared and disseminated.

Landcare is emerging as a method for empowering local government and communities to disseminate conservation farming and agro-forestry practices effectively and inexpensively. The experiences and lessons learned in Claveria provide a strong basis for scaling up to regional and national levels, and for scaling out to other municipalities. A vision for the development of national Landcare movements is set out in Figure 2.

Currently, we are employing different models for scaling up the Landcare approach and comparing them. These are integrating the Landcare approach through:

- the regular extension programme of the municipal agriculture offices and line agencies, such as the Department of Agrarian Reform and the Department of Environment and Natural Resources;
- government special projects;
- NGO development programmes;
- special bodies, such as the Cagayan-Iligan Corridor Watershed Management Council;
- watershed management and development planning of the municipality and province.

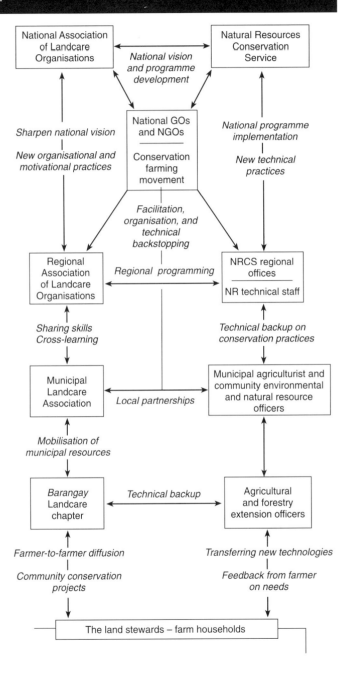

Figure 2: Conceptual framework for the vertical scaling up of the Landcare approach

National Association of Landcare Organisations

Natural Resources Conservation Service

National vision and programme development

National GOs and NGOs
___
Conservation farming movement

Sharpen national vision

New organisational and motivational practices

National programme implementation

New technical practices

Facilitation, organisation, and technical backstopping

Regional Association of Landcare Organisations

Regional programming

NRCS regional offices
___
NR technical staff

Sharing skills Cross-learning

Technical backup on conservation practices

Municipal Landcare Association

Local partnerships

Municipal agriculturist and community environmental and natural resource officers

Mobilisation of municipal resources

Barangay Landcare chapter

Technical backup

Agricultural and forestry extension officers

Farmer-to-farmer diffusion

Community conservation projects

Transferring new technologies

Feedback from farmer on needs

The land stewards – farm households

The adjacent municipality of Malitbog in Bukidnon Province approached the Claveria team to assist it in developing Landcare activities. Farmer cross-site visits and training activities were arranged. An ICRAF field extension staff member was posted to Malitbog, and the local government formed a conservation team to help start Landcare activities in four pilot *barangays* (Saguinhon 1998). Municipal funds were provided to assist Landcare chapters to establish nurseries, to fund training and cross-site visits, and to provide transport and allowance for the participants who attend monthly meetings. Based on specific requests, various study tours and training activities were organised for farmers, NGOs, and local government units interested in the Landcare approach. The ICRAF–Lantapan team has also applied the Landcare principles and approach to its work on decentralised planning and implementation of natural resource management. It helped develop a farmer agroforestry tree seed association. The movement grew to over 60 farmer groups in Lantapan and has spread to several other municipalities in central and southern Mindanao and in the Visasyan islands of Bohol and Leyte.

Many local governments, and the NGOs that are supporting rural development and environment programmes, have approached the Landcare programme to learn how to encourage it in their areas. The Department of Agrarian Reform and the Department of Environment and Natural Resources are keen to infuse the Landcare approach into their work throughout the Philippines. This has prompted ICRAF to develop a training-of-trainers strategy and methodology to accelerate the training of Landcare facilitators in government agencies and NGOs.

The new Philippines National Strategy for Improved Watershed Resources Management (DENR 1998) has incorporated the Landcare approach into its key institutional elements and operational framework. As the strategy moves into the implementation phase, it provides a good opportunity to scale up useful Landcare principles and experiences in other parts of the Philippines. However, this scaling-up process must respect and adhere to the critical, underlying elements, such as farmer voluntary action and local government partnership, that have made Landcare successful.

The term 'Landcare' originated in Australia, where a Landcare movement that has evolved since the late 1980s now encompasses over 4500 groups nationwide (Campbell and Siepen 1994). The Philippine Landcare movement adopted the same name, although it evolved independently. There is now strong interaction and exchange between the Landcare movements in Australia and those in the Philippines.

We see the prospect for research and development to be carried out through Landcare groups and to be managed by them. This would multiply the amount of work and the diversity of trials that can be accomplished, and ensure a robust understanding of the performance and recommendation domain of technical innovations. Currently, we are conducting surveys through the Landcare groups to get grassroots feedback on the priorities for research, from the farmers' perspective. In Australia, public-sector research institutions such as the Commonwealth Scientific and Industrial Research Organisation (CSIRO) are adjusting to the new reality that, through Landcare, farmers sit on the boards that decide on research project funding, and may even dominate them. This is having a galvanising effect in focusing researchers on problems that are of concern to farmers.

We may summarise by listing four functions of farmer-led, knowledge-sharing Landcare organisations:

- enhanced efficiency of extension and diffusion of improved practices (more cost effective than conventional extension functions);
- community-scale searching process for new solutions or adaptations, suited to the diverse and complex environments of smallholder farming;
- enhanced research through engagement by large numbers of smallholders in formal and informal tests of new practices;
- mobilisation in the community to understand and address landscape-level environmental problems related to water quality, forest and biodiversity protection, soil conservation, and others.

There are three significant concerns about the sustainability of the Landcare movement. Firstly, the Landcare concept is sufficiently popular that there is a definite risk of attracting support projects that do not understand the concept and that provide funds in a top-down, target-driven mode that defeats the whole basis of a farmer-led movement. The second concern is the question of how such movements can sustain themselves in the long run. Networking, and stimulation from outside contacts, is widely considered to be crucial in the long-term success of such institutions. This can be provided through Landcare Federations, which have evolved locally in Claveria, and through provincial and national federations, currently being explored in the Philippines. Thirdly, group leadership is a time-consuming and exhausting task, particularly when it is undertaken on a voluntary basis. Landcare is still very young in both the Philippines and Australia, but increasingly leadership burn-out is discussed as a concern.

Our analysis indicates that the following actions need to be taken in order to release the power of the Landcare concept further. The public sector and the non-governmental sector can help to form groups and networks, enabling them to grow, developing their managerial capabilities, and enhancing their ability to capture new information from the outside world. They can also provide leadership training to farmer leaders, helping ensure the sustainability of the organisations. Cost-sharing external assistance can also be provided. For this, the use of trust funds should be emphasised, where farmer groups can compete for small grants to implement their own local Landcare projects. This has been remarkably successful in the Australian Landcare movement. We envisage that the Landcare approach may be suited to other locations in the Philippines and elsewhere, providing a national focus for farmers to sustain the management of their resources with minimal local government support.

References

Agus, Fahmudin (1993) 'Soil Processes and Crop Production under Contour Hedgerow Systems on Sloping Oxisols', PhD dissertation, Raleigh NC: North Carolina State University

Campbell, A. and G. Siepen (1994) *Landcare: Communities Shaping the Land and the Future*, Sydney: Allen & Unwin

Cenas, P.A. and S. Pandey (1995) 'Contour Hedgerow Technology in Claveria, Misamis Oriental', paper presented at the Federation of Crop Science Society of the Philippines held at Siliman University, Dumaguete City

Department of Environment Natural Resources (DENR) (1998) *The Philippine National Strategy for Improved Watershed Resources Management*, Manila: DENR

Fujisaka, Samuel (1989) 'The need to build upon farmer practice and knowledge reminders from selected upland conservation projects and policies', *Agroforestry Systems* 9: 141–53

Fujisaka, Samuel, E. Jayson, and A. Dapusala (1994) 'Trees, grasses, and weeds: species choices in farmer-developed contour hedgerows', *Agroforestry Systems* 25: 13–32

Garrity, Dennis P. (1996) 'Conservation Tillage: Southeast Asian Perspective', paper presented at Conservation Tillage Workshop, Los Baños, Philippines, 11–12 November

Garrity, Dennis P. and P.A. Agustin (1994) 'Historical land use evolution in a tropical upland agroecosystem', *Agriculture, Ecosystems and Environment* 53: 83–95

Garrity, Dennis P. and Agustin Mercado, Jr. (1998) 'The Landcare Approach: A Two-Pronged Method to Rapidly Disseminate Agroforestry Practices in Upland Watersheds', Bogor, Indonesia: International Centre for Research in Agroforestry, Southeast Asian Regional Research Programme

International Centre for Research in Agroforestry (ICRAF) (1996) *Annual Report 1996*, Nairobi: ICRAF

Magcale-Macandog, D.B., Ken Menz, P. Rocamora, and C. Predo (1997) 'Smallholder Timber Production and Marketing: The Case of *Gmelina arborea* in Claveria, Northern Mindanao, Philippines', unpublished paper, Los Baños, Philippines: Southeast Asian Regional Centre for Graduate Study and Research in Agriculture (SEARCA)

Mercado, Agustin R., Dennis P. Garrity, Nestor Sanchez, and L. Laput (1997) 'Effect of Natural Vegetative Filter Strips Density on Crop Production and Soil Loss', paper presented at the 13th Annual Scientific Conference of the Federation of Crop Science Societies of the Philippines, Baguio City, Philippines

Saguinhon, Judith (1998) 'Scaling-up of Landcare in Malitbog, Bukidnon', paper presented during the Vietnam/Philippine Roving Workshop on Conservation Farming on Sloping Lands, Cagayan de Oro City, Philippines, 1–8 November

Stark, Marco (1999) 'Soil Management Strategies to Sustain Continuous Crop Production Between Vegetative Contour Strips on Humid Tropical Hillsides: Technology Development and Dissemination based on Farmers' Adaptive Field Experimentation in the Philippines', PhD dissertation, Witzenhausen, Germany: University of Kassel

Stark, Marco, Dennis P. Garrity, Agustin Mercado, Jr., and Samuel C. Jutzi (2000) 'Building Research on Farmers' Innovations: Soil Conservation using Low-Cost Natural Vegetative Filter Strips', paper presented at the Environmental Education Network of the Philippines, Misamis Oriental State College of Agriculture, Claveria, 31 May–1 June

Swete-Kelly, David E. (1997) 'Systems for steep lands bean production', in J. Hanna (ed.) *Landcare: Best Practice*, Canberra: National Heritage Trust

Scaling up adoption and impact of agroforestry technologies:
experiences from western Kenya

Qureish Noordin, Amadou Niang, Bashir Jama, and Mary Nyasimi

Western Kenya, a densely populated region of the country, is an example of many areas in Africa where the continued threat to the world's land resources is compounded by the need to raise food production and reduce poverty. Here, attainment of food security is intrinsically linked with reversing agricultural stagnation, safeguarding the natural resource base, slowing population growth rates, combating the negative impacts of the HIV/AIDS pandemic on the community, and reducing poverty.

Farmers in this region, with farm size typically less than 1 ha per household, have many problems. Key among these are low and declining soil fertility, which is reflected in low crop yield (maize yields being typically less than 1 tonne grain per hectare); fodder and fuelwood shortages; and low incomes from farming activities. Important consequences of these problems include widespread poverty – over half of the households in the region live in absolute poverty, below the World Bank's figure of US$1 a day; severe food insecurity – many families produce little or no food during three to nine months a year; high rural-to-urban migration; and high environmental degradation, including Lake Victoria.

Scope of the paper

Over the last seven years, the International Centre for Research in Agroforestry (ICRAF) and its national collaborators in western Kenya, the Kenya Forestry Research Institute (KEFRI), and the Kenya Agricultural Research Institute (KARI), have been evaluating and disseminating several agroforestry technologies for improving farm productivity and incomes. Farmers and communities have been key participants in this research–development continuum. Several options and adaptations for soil fertility management and conservation have been developed. Examples include short-duration improved fallows with fast-growing leguminous trees and shrubs, and biomass transfer of *Tithonia diversfolia* – the leafy biomass cut from hedges on farm boundaries and from roadsides and spread on crop fields. These practices provide ample quantities of nitrogen to the soil. Their integration with phosphorus fertilisers (including phosphate rock found in the region) is an effective and economically feasible means to improve soil fertility and productivity (Swinkels *et al.* 1997; Jama *et al.* 1998, 2000; De Wolf *et al.* 2000). In addition to improving soil fertility, several of the species used for improved fallows provide fuelwood and stakes for supporting crops such as tomato and climbing bean. Planting *Tithonia* and *Calliandra calothyrsus* as dense hedges on contour lines has also become an attractive option for soil conservation, and *Calliandra* also has fodder value. Many species that provide fuelwood and timber, such as *Grevillea robusta*, have also been disseminated.

To disseminate the technologies available, in 1997 we initiated a pilot project for testing approaches. Key challenges to be addressed included the question of how to bridge the information and knowledge gap between research and farmers that was responsible for the low and declining agricultural productivity and increasing poverty of the farmers (Niang *et al.* 1999). Two approaches were examined: (1) establishing pilot projects and sites that promote the use of community or village-based organisations such as women's, church, and youth groups, and (2) facilitating the extension service and other development partners.

The pilot project we describe here builds on the experience of two development projects – CARE-Kenya and KWAP (Kenya Woodfuel and Agroforestry Project) – that have used community-based approaches to disseminate agroforestry technologies in western Kenya.

The KARI/KEFRI/ICRAF pilot project approach: making dissemination a community responsibility

Dissemination on a wide scale is complicated by several factors. First, much learning and interaction are required to introduce improved fallows as a biological system, because they are not simple adjustments in current or past farming practices (Place *et al.* 2000). Second, while pilot projects enhance adoption and impact, they cannot be replicated everywhere. Third, extension services are weak in many African countries because financial support for them is poor.

In Kenya, the government extension service has traditionally been the main method of disseminating agricultural technology to farmers. However, given the retrenchment programme in progress and the limited resources in funds and materials, including logistics, its impact has been small or at best very localised. Its limited transport resources mean that the extension service has focused on using contact farmers to reach other farmers. It soon became apparent, however, that the contact farmers selected are typically well off, and often do not represent the poor, which is the most important target group. This can lead to limited transfer of information and technologies to the target farmers. To mitigate these problems, several projects have attempted approaches to improve upon the existing limited extension services by engaging community-based organisations (CBOs). Here, we highlight the two major agroforestry projects upon which the pilot project was based.

Kenya Woodfuel and Agroforestry Project

The Kenya Woodfuel and Agroforestry Project (KWAP) operated in Busia District of western Kenya between 1990 and 1997. KWAP used an A–B–C framework to implement its dissemination activities. The pilot area was A, which was a catchment; B was the administrative location in which the pilot area falls; and C was the intervention agro-ecological zone in which the pilot area falls. In area A, KWAP worked intensively with partners, including extension agents from government agencies and NGOs operating in that area. In areas B and C, KWAP left the work to line agencies with a mandate to offer extension services, and its role was to help these line agencies carry out their duties.

Farmer groups in A areas such as catchment committees, women's groups, youth groups, and adult educational groups, had different group activities. KWAP helped these groups consolidate into umbrella development groups (UDGs), to give them better bargaining power for acquiring information and resources. These UDGs were responsible for co-ordinating and steering the development activities of individual groups.

An umbrella development group comprised 50–60 members who represented the various groups existing in the catchment. Each group had 25–30 members, and each selected two members to represent it in the UDG. Each UDG had various subcommittees with different responsibilities, for instance the adaptive research farmers' committee, whose role was to develop and test any promising technology on behalf of the community. All UDG members were resource persons for their respective groups in the technology transfer process. The whole programme covered six catchments, with six to eight adaptive research farmers in each catchment. Every resource person had three or four follower farmers for closer guidance in their respective groups. KWAP's role in this farmer-to-farmer information exchange was to strengthen the UDGs in their technical and managerial capacities through training and educational tours for farmers.

In these UDGs, farmers took centre stage in all their developmental activities, which initiated research. As the UDGs worked with the farmer groups, structural weaknesses emerged that needed to be addressed to make them more effective in handling their agricultural development activities. The main ones were the lack of institutional support for participating farmer organisations in knowledge, resources, and logistics once the supporting NGO wound up its activities in that district; insufficient skills in conflict resolution and record keeping; and lack of knowledge about the adaptive research process.

CARE Agroforestry Project

In adjacent Siaya District, CARE's Agroforestry Project also facilitated a community-based approach for the ten years from 1988 to 1998. In this project, CARE worked with women's groups and schools in 20 locations. In each location, every group of 15–20 members had selected four or five group resource persons (GRPs) who were knowledgeable and were able to disseminate technical messages. In each GRP was one adaptive research farmer who conducted trials on behalf of the group members. One CARE extension staff person provided technical backup for 12 GRPs. In the location, a co-ordinating committee, known as the locational agroforestry committee, comprised representatives of adaptive research farmers, GRPs, government extension staff, and the provincial administration.

Using this approach, target farms did well in tree planting, and a vital link between farmers and researchers was developed. However, the groups had little input into the choice of technologies. Disseminating information to other group members was passive and slow, leading to a insignificant multiplier effect. The approach was also top-down, and lacked the support of village groups and organisations at the grassroots.

The KARI/KEFRI/ICRAF pilot project village committee approach

Building on CARE's experiences, the KARI/KEFRI/ICRAF pilot project in Siaya District in 1997 engaged a *village approach*. Specifically, this approach aims to make all farmers in an entire village become adaptive research farmers by working with groups that are representative of village committees as a means of creating awareness and disseminating information and technologies on a wide scale.

The purpose of the village committee was, therefore, to disseminate technologies to all farmers. This is intended to lead to the active and full participation of all community members to ensure that use of a technology would continue after a project ended. The method relies on using existing village organisational structures. In most villages in western Kenya, organised groups formed for various purposes exist – commonly church groups, women's and youth self-help groups, and clan and sub-clan organisations. Groups include a mix of farmers, including men and women of all ages, ethnicities, and degrees of wealth with different needs, constraints, and opportunities. Groups vary in size from 15 or 20 members to entire clans in a village. Though membership may spread to other villages, it is usually contained within the village in question.

Typically, a village contains between 80 and 140 households, a sub-location contains 240–320 households, and a location contains 680–750 households or 4–5000 people. Extension agents, who are based in the location, can pass information to farmers through the sub-location and village committees, who pass it to the farmers through village groups and social organisations (see Figure 1).

The way in which the village committees are formed in the project is described in detail in Niang *et al.* (1999) and Noordin *et al.* (2000). In brief, the main task was to determine, through consultative meetings with groups of farmers, a structure that would facilitate widespread dissemination of the technologies to all farmers in a participatory manner. One approach that looks promising is to form committees in the village, the sub-location, and the location.

Forming the village committees starts with identifying all the groups that exist in a given village. All such groups, large or small, are represented equally. Members of each group use their own criteria to select a delegate to a village committee, which is made up of the farmers thus selected. The village committee selects two delegates to represent their village in the sub-location.

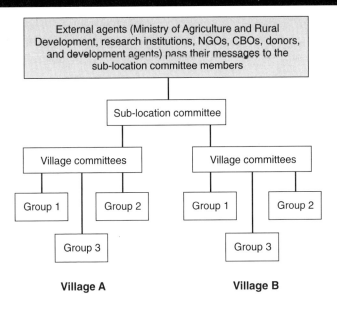

Village A   **Village B**

Using this process, we formed over 28 village committees and five sub-locational committees in a short period, the latter being made up of farmers from between two and five villages.

We adopted this approach for all villages in the pilot area. Through the committees, external facilitators or development agents now have entry points to understand village problems. Likewise, the committees provide a way to communicate beyond the village and to disseminate new ideas to village households. Existing informal organisations provide great potential for building on what works rather than creating new structures that lack proper foundation, and which are bound to collapse. Groups also help in changing attitudes of members, especially in dispelling taboos and myths that might relate to certain trees or farming practices. The poor also belong to groups, particularly church-related ones. According to one study in the pilot villages, over 80 per cent of the members of church groups belong to the poorest category of the village community (Mary Nyasimi 2000, unpublished data). Working with such groups ensures that women-led households and the poor in the villages are effectively reached.

Once formed, the committees were trained in several areas necessary to improve their performance. Key among these were:

- technical aspects of agroforestry – seed production, handling, and storage; nursery establishment and management; soil conservation and soil fertility replenishment innovations; management of high-value trees;
- group dynamics and team building;
- record keeping;
- leadership skills, since members represent fellow farmers in meetings;
- monitoring and evaluation of projects; and
- proposal writing.

Achievements and impacts made with the village approach

The village committees helped to mobilise collective action for activities such as soil and water conservation that were agreed upon by all as the starting point for a sound soil fertility management programme. Farmers made contour bunds of *Tithonia* hedges on their farms and villages and occasionally across villages. This activity required the input of the extension staff and the co-operation of the farmers. The village committees ensured that this happened. The extension staff needed assistance, particularly for transport; through re-training, they gained the necessary skills to facilitate farmer participation.

Within a short period, community participation led to wide-scale on-farm testing and uptake of improved fallows and biomass transfer. This was partly because farmers found the technology attractive. Fertilisers are expensive, and many farmers cannot afford them. But they needed no money for improved fallow seeds or for harvesting the *Tithonia* already present in their villages. Consequently, an impressive feature of the technologies is that they are being used by the poor and by women. A recent study using quantitative (logit) analysis of over 1100 households found that while wealth was positively related to the use of fertiliser, compost, and manure, it was not related to the use of improved fallows or biomass transfer (Place *et al.* 2000). Similarly, female-headed households are less likely than male-headed ones to use fertiliser, but they are equally likely to use the agroforestry systems. Within the pilot project area, monitoring and evaluation exercises indicate that farmers using improved fallows have increased the average size of fallow from 134 to 247 m² between 1997 and 1999 (Pisanelli and Franzel 1999).

Within communities, initial results from interviews with participating farmers and village elders indicate that improvement has been marked in food availability during traditional food-deficit periods (the period following the short rains when very little maize is produced because of erratic rainfall and associated high incidence of crop pests and diseases). The elders and farmers also report that pilferage of maize in the field has lessened drastically because villagers have an adequate supply of food. Some farmers have also shifted to growing higher-value crops like kale, carrot, tomato, and onion, which they can do because the fertility of their soil has improved.

To support income-generation initiatives of farmers, the project introduced several high-value timber, fruit, and medicinal tree species. The community identified individuals and groups that could undertake the task of multiplying the seed and planting materials required. For example, over 50 farmers established bulking plots and mother blocks of high-value mango and avocado trees on behalf on their villages. Also, volunteer farmers established over 30 community seed stands of improved fallow. Unfortunately, although the entire village is supposed to manage the seed stands through the various social groups, often only a few farmers or only the farmer on whose farm the seed stand is located end up performing this task.

Integrating inorganic phosphorus with organic options such as improved fallows and biomass transfer is essential for enhancing yields and adoption of the agroforestry technologies. In addition to commercially available phosphorus fertilisers, we promoted use of a reactive phosphate rock. This was facilitated through a pilot credit scheme run by women's and youth groups in 19 villages. Over two consecutive years, repayment rates from farmers to village committees (and then back to the project) have been encouraging – on average, 64 per cent after three seasons, and similar for both men and women.

The pilot villages are now acting as focal training points for farmers in other Kenyan villages, and also in Ugandan and Tanzanian villages. They have become a valuable way of linking up with development partners who are also involved in scaling up agroforestry and other agricultural innovations.

Lessons learned and challenges ahead

Village committees can be an effective means of disseminating technology through creating awareness and following up with their members. Groups create cohesiveness and togetherness among community members. Using existing groups (rather than forming new ones) accelerates and enhances impact. It empowers the groups and gives them a sense of ownership over the development process. Mama Dorcas, chairperson of a women's group of Vihiga District, emphasises the value and gains to be made by using existing groups. Group members will follow up and disseminate the technology both to other members and to non-members. Awareness creation should take the form of mass campaigns, using all avenues possible – churches, schools, public gatherings, farm-to-farm visits, and so on. Mama Dorcas says that this is how the family planning campaigns succeeded in her area in the mid-1980s and that we have to do the same for agro-forestry for it to succeed also. She emphasises the need to focus on women since women perform nearly all the farm work in her area.

In general, individual groups (particularly women's groups) were more active than the groups forming the village committees. Many groups, however, remain inactive. This is particularly true of those whose formation was associated with gains to be made from political events such as national or local elections.

The lack of adequate funds to conduct activities is another reason for group inactivity. The expectation of groups and farmers in general is that they will receive some financial support from the project. We have avoided this situation and consequently have ended up with active, self-supporting groups. In the process, we have helped some groups to develop proposals and get funding. However, when we work more with certain groups, this can generate a wider perception of partiality on our part.

The sub-location and location committees are the weakest group in this chain of grassroots organisations. Developing village action plans is one of the key functions the village committees were expected to perform but in most villages they have not done so. Village committee elders held on to elective offices but rarely called village meetings. This situation created disillusionment and tension among members. The perception arose that the roles and responsibilities of the village committee were not clear, though it should be involved in planning and co-ordinating soil conservation activities, supplying inputs if collective action is needed, and organising study tours. Such activities require frequent follow-up.

We found that project field officers needed to follow up and encourage the village committees and farmers persistently. Where follow-up was weak, so were the uptake of the technologies and the performance of the group. Surprisingly, this was so even after three years of interaction with some of the groups. Farmers always requested exchange and study tours and benefited from them. But key challenges with the study tours concerned the questions of who got to go, how they were selected, and how to ensure that information and materials such as seeds obtained on the study tour reached those who did not participate. Often this sharing did not happen, creating envy and enmity among village members and between villagers and the project field officers.

Farmers often describe field exchange visits as a real eye-opener and an inspiration to those who participate. The exchange visits create networking among farmers. The main limitation is that visits can be expensive, depending on the distance travelled and the number of farmers involved. Typically only one tour is conducted per village, and farmers reckon this is not enough. A tour also requires follow-up to see if what was learned is being used and whether the message spread beyond those who took part in the tour. Often it does not, and so the gains made from the visit are few and remain virtually unknown beyond the fortunate few who toured.

To facilitate the scaling-up efforts by CBOs, NGOs, and farmers, KARI in 2000 initiated the Agricultural Technology and Information Response Initiative (ATIRI). This initiative is a competitive grant mechanism that aims to strengthen the link between research and extension. Already some of the groups involved in the KARI/ KEFRI/ICRAF pilot project have received funding from ATIRI after writing successful proposals.

Scaling up through the activities of other development partners

Collaborating with many organisations, specifically those focusing on issues of soil fertility management, provided us with the opportunity to scale up the lessons learned from the pilot project. The main institutions and projects with which we collaborated, and the way in which we scaled up, are described below.

Government ministries and projects

We worked with government extension services at the catchment level within the framework of the government Soil and Water Conservation Programme, and with the National Agricultural and Livestock Extension Programme (NALEP), whose focus is on farmer contact groups.

In this approach, extension staff are trained in participatory approaches, agroforestry interventions, tree propagation, and seed production techniques. Activities include making field visits, planning meetings, and providing extension materials. The extension staff pass information to farmers through various methods that include training-and-visit, demonstrations, and farmer field schools.

A farmer field school is a group extension method based on adult education methods. It is a 'school without walls' that teaches basic crop and livestock agro-ecology and management, making farmers experts on their own farms. It comprises groups of farmers who meet regularly during the growing season to experiment with new production options. After the training period the farmers continue to meet and share information but with less contact with extension agents. After taking part in a farmer field school, participants are able to train others in improved crop and animal husbandry, leaving extension staff free to cover other areas. A drawback is that the process can be tedious, especially when used for slow-maturing crops, as farmers must meet weekly (Kibisu and Khisa 2000).

Adaptive research farmers – the KARI–Kisii approach

The Kisii station of KARI also employs a farmer participatory approach to test and disseminate technologies (Okoko *et al.* 2000). Farmers select whom they want to participate in both research and demonstration activities. These farmers, each representing a village, then establish a farmer research committee that assists in implementing, monitoring, and evaluating the technologies. Committee members share information they acquire with the farmers of their villages through a farmer planning and evaluation workshop, demonstration and field days, and farmer exchange visits.

A Participatory Learning Action Research village project

To integrate and institutionalise participatory technology development and dissemination skills among the government extension services, we collaborated with a KARI-led, Dutch-funded pilot project, Participatory

Learning Action Research (PLAR). Through government extension, the project established seven PLAR villages in seven districts. These villages are now regarded as 'learning points'. Working with extension staff, these communities drive their own development process to identify, implement, and evaluate their development initiatives. The approach is similar to that being used in the pilot project villages – participatory decision-making, where the pilot project plays a facilitation role. The seven villages are now important satellites for training and for disseminating soil fertility management options (Gacheru *et al.* 2000).

As part of the start-up process, farmers went on study tours to villages where the technologies had been in practice for at least three years. Farmers and the extension staff gained trust in each other and mixed and discussed freely. Farmers say that PLAR has let them get acquainted with each other and see what is happening on other farms, and it has helped them understand technical messages behind the results they see. By the end of one season, farmers felt more confident in themselves because of their expanded knowledge, and their demands were for more knowledge – not inputs.

One of the important PLAR steps is to develop village action plans. These plans help farmers to set priorities and to decide how to execute them in a manner that is participatory and that involves all in the village. The plans create community ownership of the activities. Experience from several villages shows that enthusiasm is high when planning begins, but that many farmers drop out of the village action plan meetings once they realise that they will not get free inputs. Also, farmers detested keeping records of farm activities – reasoning that they could keep records in their heads. Where records were kept, they were kept by young people and schoolchildren.

To accelerate scaling up, farmers suggested that improved fallows be planted on farms near the road so that people could observe them and learn from them as they passed by, and might then want to emulate them. They also wanted to bring the fertiliser and seed input dealers closer to their village.

The African Highlands Initiative approach

The African Highlands Initiative programme in western Kenya is testing an approach in participatory technology dissemination in which five pilot villages form farmer committees to extend the adaptive research findings of trial farmers that the village community selects.

Within a village committee is a sub-committee of resource persons who work with adaptive research farmers and train other farmers in technologies already tried in the pilot villages. These sub-committees have representatives in a farmer research committee who are responsible for co-ordinating research and dissemination activities in all pilot villages. The farmer research committee links resource persons in various sub-committees in the pilot village to those in other villages so that they can train the new village committees in both the adaptive research process and the technologies. Communities outside pilot areas get their information from the farmer research committees. Through these farmer networks, we can diffuse information about the technologies and their sustainability more widely.

With this community-based system, farmers play a central role in exchanging information geared towards solving their perceived farming problems, and they help break down community-related barriers that hinder the free flow of information.

After only two years on the ground, the initiative has already achieved a lot. Five pilot villages (approximately 10,000 people) have established demonstration plots. Each village has a village committee for adaptive research and dissemination.

The Tropical Soil Biology and Fertility interactive learning project

In the same villages in which the African Highlands Initiative is working, the Tropical Soil Biology and Fertility programme is testing a community-based interactive learning approach that aims to improve agricultural productivity by incorporating scientific principles into farmers' ecological and practical knowledge. It is a strategy geared towards strengthening farmers' knowledge base. Specifically, this approach aims to achieve the following:

- to identify and document farmers' existing agro-ecological knowledge (folk ecology) and their knowledge gaps;
- to communicate new scientific knowledge to farmers to strengthen their understanding of agro-ecological and soil biological processes, including nitrogen fixation, soil organisms and decomposition, organic resource quality, and fertiliser equivalencies;
- to use innovative tools to communicate required scientific knowledge to farmers, including community laboratories, microscopes, pot experiments, nutrient test strips, posters, video, drama;

- to use different platforms of communication to reach different categories of people within communities, including churches, schools, women's groups, clan groups;
- to use appropriate forums and media to communicate farmers' agro-ecological knowledge to scientists and extension agents with the view to incorporating this local knowledge into scientific debate and research.

Little can be said at this point about the efficacy of these community-based approaches. However, they do have strong links and backup from both national and international research and development programmes working together in an integrated manner.

Non-governmental organisations

Many NGOs are operating in western Kenya, in agriculture, energy, food security, water, credit systems, farm input supplies, and forestry. Their field extension workers are trained in various technologies, and they are able to transfer the information to the farmers with whom they work in their particular mandate areas.

Among our NGO partners are CARE, the On-Farm Productivity Enhancement Programme, PLAN International, Action Aid, Hortiequip Ltd., the Organic Matter Management Network, the Rural Energy and Food Security Programme, the Vi-Agroforestry Programme, Africa Now, Care for the Earth, Community Mobilization Against Desertification, Ideas Research Management Consultants, and the Sustainable Community Oriented Development Project. Most of these NGOs, however, lack staff who are well-trained in disseminating information. Also, NGO presence often ends with their project, thus creating a lack of long-term commitment and sustainability.

CARE is an exception in that it has had long-term presence in the region. As mentioned previously, it has had a successful programme for ten years in Siaya District, north of Lake Victoria, where it has developed a community-based dissemination system using groups within administrative locations. Over the last three years, CARE has moved the programme into new districts to the south of Lake Victoria, where it now uses a participatory extension method referred to as Training Resource Persons in Agriculture for Community Extension (TRACE).

Often, farmers learn best from their peers and neighbours, adopting many agricultural innovations by learning from fellow farmers.

The TRACE process aims to establish a functional and sustainable community extension process based on resource persons; to build the capability within the community to assess and manage agricultural information and services with the aim of improving agricultural productivity; and to establish a community-based adaptive research process through which farmers are able to manage and make their own assessment of the technologies they try. The starting point is to create awareness, and then to select members of the community who will participate in the various village, sub-location, and locational institutions. This is done after the chiefs and leaders have been given orientation training.

A number of benefits are associated with TRACE:

- Community participation in decision making is ensured; decisions are reached by consensus.

- The community has a sense of ownership of the process because locational management committees take responsibility for the processes involved.

- Sustainability is ensured through building the capacity of community resource persons and management committees.

- Community involvement in monitoring and evaluation ensures improvement in the methodology.

- The process allows for close links with the government administrative system, given that the locational management committees are sub-committees of the locational development committees of districts and therefore receive some funding from the government.

- Geographical coverage is wide – the unit of operation is a location with many villages and people instead of individual villages; and the groups and villages within a location are evenly distributed, ensuring that a large proportion of the people are reached.

The success of this and other hierarchical organisational schemes depends on the rate of information from the top (that is, from the locational development committees) to the villages and the groups within them. Because the committees lose motivation and sense of purpose if there is no new injection of information and innovations, they must establish external links with other agencies to ensure that the process continues.

Community-based organisations

A number of CBOs also work in the area. They include the Community Organic Farming Development Organisation, Ugunja Community Resource Centre, Rachuonyo Youth Skills Development Programme, the Locational Agroforestry Committee – Kanyaluo, Sacred-Africa, Sustainable Community Environmental Programme, and the Maseno Inter-Christian Child Welfare Programme. Such CBOs can be particularly effective in building capacity within the community, achieving wider geographical coverage, and ensuring continuity and sustainability of activities after donor-funded projects end. They afford larger organisations the opportunity to reach farmers more easily. They also provide effective feedback. Most, however, lack adequate operating resources, skilled staff, and good leadership.

Educational institutions

Schools provide a good forum for passing urgent messages to community members within a short time (Noordin 1996). Schools not only act as an effective medium of communication but also function as facilitators for a given intervention. Through demonstrations at the schools, we aim to reach the community directly or indirectly, and parents are able to discuss and evaluate these demonstrations during parents' days and even in normal school days. The community uses the school for bulking plots, particularly for producing improved fallow seed. Clubs within the school may raise seedlings of high-value trees, which are planted either in the school compound or at club members' homes. This approach effectively prepares the children as future farmers who will put what they have learned into practice.

Churches and social groups

Churches and social groups have been found to be better than individual or contact farmers as entry points to extension in a community (Mungala and Chavangi 1996) and have been used by many organisations. At Maseno, group contacts for women's and youth groups interested in agroforestry innovations disseminate information and, through them, more farmers are reached. In total, 23 youth and women's groups are working directly with the researchers to disseminate agroforestry messages. However, illiteracy and socio-cultural barriers are hurdles that at times prove difficult to surmount.

Lessons learned and challenges ahead using development partners

The challenges have been many in leading and maintaining a large number of development partners to scale up agroforestry. Key among them have been:

- high transaction costs (staff time and operational funds);
- lack of commitment in the absence of joint resources and memoranda of understanding;
- obstacles in pooling resources, leading to competition and duplication of effort;
- weak links in the researcher–extension–farmer continuum;
- weak documentation of the research activities conducted in the region, hence little exchange of information among the various stakeholders;
- lack of sustainability among NGOs, with short-lived projects leading to lack of continuity or long-term commitment; and
- lack of operational funds for the mainstream extension services of the government and unexpected transfers of field staff, leading to interruption of planned activities.

Institutionalising and strengthening partnerships

The main task ahead of the programme now is scaling up and spreading the benefits of agroforestry out beyond the pilot villages to the six million or so potential smallholder farmers in western Kenya. Towards this, a Consortium for Increasing Farm Productivity in Western Kenya was launched in January 2001.

This consortium contains over 40 organisations involved in agriculture, including agroforestry research and development, research and development organisations such as KARI and KEFRI, the extension branch of the Ministry of Agriculture and Rural Development, NGOs, CBOs, the Regional Land Management Unit (RELMA) of the Swedish International Development Co-operation Agency (SIDA), and farmer groups and associations. Its co-ordination committee represents ten institutions, including local councils, local representatives of the HIV/AIDS programme of the Ministry of Health, and the Forest Department. This arrangement will allow the pilot project to operate in a larger number of locations and effectively cover the 20 districts in western Kenya.

The consortium will create a forum for greater commitment, complementarity, and networking among partners. As a starting point, the first workshop in 2000 documented the various technological options and the methods that different partners use to disseminate them (Nyasimi *et al.* 2000).

To back up the consortium, the KARI/KEFRI/ICRAF pilot project in western Kenya will strengthen and continue to provide the following services:

- Training of development agents in participatory methods and technical aspects. At the same time, training farmers in the partners' mandate areas.
- Creating awareness through field days, visits, and tours for the partners.
- Attendance at the annual agricultural shows in various locations.
- Production of extension and training materials for partners and farmers.
- Establishment of seed production stands with partners.
- Organisation of regular joint planning meetings.
- Production of the quarterly newsletter *Miti Ni Maendeleo*, meaning 'Trees for Development', presently published jointly with the GTZ-supported project Integration of Trees into Farming Systems.

The challenge that now remains is to put the plan into operation to achieve the desired objectives.

Conclusions and future needs

Scaling up agroforestry technologies means creating awareness, training farmers, and encouraging participation of the community at large. Towards this objective, the projects and partnership existing in western Kenya have engaged in slightly different approaches but all with a common theme and a strong focus on CBOs, such as women's, youth, and church groups. These approaches present strong evidence that CBOs have great potential to empower community members to become their own agents of change and that they can bring farmers closer to government institutions and other service providers such as microcredit institutions and research and development organisations.

To achieve reasonable community development, community members should articulate their problems well and even suggest home-grown solutions. Enlightened farmers can make their own decisions when they are well informed. Such a scenario can be attained when communication is a two-way channel between the farmers and the researchers and extension agents. For farmers to handle community developmental activities effectively, their leaders need to be equipped with both leadership and management skills.

Consistent follow-up and support from projects and development agencies seems crucial to the performance of CBOs. We found that where follow-up was weak, uptake of the technologies and performance of the group was also generally weak. A key challenge, then, is in sustaining this follow-up, and particularly in addressing how either the mainstream extension service or NGOs and CBOs can do this once the project ends.

To scale up beyond pilot sites into larger geographical regions, it is essential for partners engaged in similar activities to collaborate and co-operate. Doing so minimises duplication and competition. It creates synergy, adds value, and enhances impact. It is for this reason that we have invested energy and resources in forming the Agroforestry Consortium for Western Kenya. The remaining challenge is to make it deliver in a cost-effective manner that is also sustainable after project resources are withdrawn.

Acknowledgements

A number of researchers in western Kenya have co-operated in carrying out this scaling-up work. Among those involved are Stephen Ruigu and Aggrey Otieno, who work for ICRAF; Eva Gacheru, John Ojiem, and Daniel Rotich with the Kenya Agricultural Research Institute; John Mukalama and Issack Ekise with the Tropical Soil Biology and Fertility programme; George Etindi with the Kenya Forestry Research Institute; Electine Wabwile and Godrick Khisa with the Kenya Ministry of Agriculture and Rural Development; and Loice Omoro with CARE-Kenya. Thanks go also to the government of Kenya for financial and staff support, and to the Rockefeller Foundation, European Union, and the Swedish International Development Co-operation Agency (SIDA) for their financial support.

References

De Wolf, J., R. Rommelse, and A. Pisanelli (2000) 'Improved Fallow Technology in Western Kenya: Potential and Reception by Farmers', Nairobi: International Centre for Research in Agroforestry (photocopy)

Gacheru, E., B. Gerard, and M. Koijman (2000) 'Participatory Learning Action Research for Integrated Soil Fertility Management: Reports from PLAR Teams in Western Kenya – Experiences from 7 Districts', Nairobi: Kenya Agricultural Research Institute

Jama, B.A., R.J. Buresh, and F. Place (1998) '*Sesbania* tree fallows on phosphorus-deficient sites: maize yields and financial benefit', *Agronomy Journal* 90: 717–26

Jama, B.A., C.A. Palm, R.J. Buresh, A.I. Niang, C. Gachengo, G. Nziguheba, and B. Amadalo (2000) '*Tithonia diversifolia* as a green manure for soil fertility improvement in western Kenya: a review', *Agroforestry Systems* 49: 201–21

Kibisu, L. and G. Khisa (2000) 'Farmer field schools', in M. Nyasimi, Q. Noordin, B. Jama, and S. Ruigu (eds) (2000)

Mungala, P. and N. Chavangi (1996) 'An overview of agroforestry extension in Kenya', in J.O. Mugo (ed.) *People and Institutional Participation in Agroforestry for Sustainable Development*, First Kenya Agroforestry Conference, held at Kenya Forestry Research Institute, 25–29 March 1996, Nairobi: Kenya Agricultural Research Institute

Niang, A., J. De Wolf, M. Nyasimi, T. Hansen, R. Rommelse, and K. Mwendwa (1999) *Soil Fertility Recapitalization and Replenishment Project in Western Kenya*, progress report, February 1997–July 1998, Pilot Project Report No. 9, Regional Agroforestry Research Centre, Maseno, Kenya, Nairobi: International Centre for Research in Agroforestry

Noordin, Q. (1996) 'Community participation in agroforestry development and extension: experience of the Kenya Woodfuel and Programme (KWAP), Busia District, Kenya', *East African Agricultural and Forestry Journal* 62(2): 261–70

Noordin, Q., M. Nyasimi, A. Niang, S. Ruigu, and B. Jama (2000) 'Facilitating dissemination and scaling up strategies in the Maseno pilot project on soil fertility replenishment and recapitalization', in M. Nyasimi, Q. Noordin, B. Jama, and S. Ruigu (eds) (2000)

Nyasimi, M., Q. Noordin, B. Jama, and S. Ruigu (eds) (2000) *Dissemination and Extension Methodologies for Integrated Soil Fertility Management Practices in Western Kenya*, proceedings of a workshop held at ICRAF, Kisumu, 25–26 January 2000, Nairobi: International Centre for Research in Agroforestry

Okoko, N., N. Kidula, C. Muyonga, and S. Obaga (2000) 'Dissemination strategies of various technologies developed by a soil management project in Kisii', in M. Nyasimi, Q. Noordin, B. Jama and S. Ruigu (eds) (2000)

Pisanelli, A. and S. Franzel (2000) 'Adoption of Improved Tree Fallows in Western Kenya: Farmer Practices, Knowledge, and Perception', Nairobi: International Centre for Research in Agroforestry (photocopy)

Place, F., S. Franzel, J. De Wolf, R. Rommelse, F.R. Kwesiga, A.I. Niang, and B.A. Jama (2000) 'Agroforestry for Soil Fertility Replenishment: Evidence on Adoption Processes in Kenya and Zambia', paper presented at a workshop 'Understanding Adoption Processes for Natural Resource Management Practices in sub-Saharan Africa', International Centre for Research in Agroforestry, Nairobi, 3–5 July 2000

Swinkels, R.A., S. Franzel, K.D. Shepherd, E. Ohlsson, and J.K. Ndufa (1997) 'The economics of short rotation improved fallows: evidence from areas of high population density in western Kenya', *Agricultural Systems* 55: 99–121

Scaling up the benefits of agroforestry research:
lessons learned and research challenges

Steven Franzel, Peter Cooper, and Glenn L. Denning

Research and development institutions are becoming increasingly committed to scaling up the adoption and impact of technical, institutional, and policy innovations that improve household livelihoods. Scaling up is a complex subject; Uvin and Miller (1999) developed a taxonomy and arrived at 17 different kinds of scaling up, focusing on structure (when a programme expands its size), strategy (degree of political involvement), and resource base (organisational strength). The International Institute of Rural Reconstruction (IIRR) (2000) gives a useful and succinct functional definition of scaling up: efforts to 'bring more quality benefits to more people over a wider geographical area more quickly, more equitably, and more lastingly'. Different users of the term consider different issues as important. Proponents of the technology-transfer paradigm often imply that the main issue in scaling up is to replicate the use of improved practices – for example, more farmers using mineral fertiliser – and they focus on such issues as delivery of inputs and demonstration of benefits (Quiñones and Gebre 1996). Others, such as Krishna *et al.* (1998), consider scaling up in much broader terms, that is, as a process of adaptation, innovation, feedback, and expanded human capability. In line with the latter approach, Cooper and Denning (2000) identified ten essential and generic elements of a successful scaling-up strategy, as noted in Denning (2001). Our paper summarises the main lessons learned from the case studies that appear in this volume and presents them by element.

Technical options

Most of the case studies involved scaling up the use of technical options that had first been developed by researchers and farmers conducting participatory research. All involved offering farmers a range of options.

In Southern Africa, for example, several species and practices were available for producing fodder trees, fuelwood, and fruit, and for enhancing soil fertility. Offering farmers alternative practices and tree species to solve a particular problem was important for several reasons:

- Farmers want to diversify income and thus reduce the multiple risks they face. For example, a single option may, over time or through expanded use, succumb to pests or diseases. Farmers also face the risk of market failure and the risks associated with season-to-season variation; thus they value multiple options.

- Different farmers are likely to have different preferences. Anyonge *et al.* (2001) found that farmers in densely populated areas preferred *Grevillea robusta* for timber, because it competed little with their crops. Farmers in sparsely populated areas, however, where farm size was larger, preferred the more competitive and faster-growing *Eucalyptus* spp..

- Different options are likely to perform differently in different environments. Weber *et al.* (2001) noted variation in ranking in wood density among provenances of a timber tree, *Calycophyllum spruceanum*, in different areas of a watershed in Peru. The variation in ranking was associated with differences in soil type and rainfall.

- Promoting different species and different provenances or varieties of the same species enhances biodiversity.

- Diversity of tree species can diversify income and thus reduce the risk of market failure.

Practices that could be adapted to a range of different biophysical and socio-economic circumstances were also useful in the scaling-up process. For example, improved fallow options in Southern Africa included a range of species that could be planted by direct seeding or by growing seedlings in nurseries, and they could be planted in pure stands or inter-cropped with maize. In addition, the different species offered different by-products, including pesticide, food, and wood for fuel and construction.

Another critical function in the case studies was defining the recommendation domains of options, that is, the biophysical and socio-economic circumstances under which farmers would adopt them. Wambugu *et al.* (2001) found that *Calliandra calothyrsus*, a fodder tree, performed poorly on acidic soils in central Kenya. Furthermore, it was not attractive in irrigated areas, where farmers preferred to use their labour to produce vegetables.

Farmer-centred research and extension

Farmer-centred research was key for generating appropriate practices for farmers and for responding to farmers' problems during the scaling-up process. Diagnostic surveys helped identify farmer problems and opportunities; farmer preference surveys and market assessments helped researchers in Peru to set priorities on species for research (Weber *et al.* 2001). In Mexico, farmers held workshops at which they selected the practices they wanted to test. Haggar *et al.* (2001) helped them form research groups and conduct their own experiments, which facilitated the exchange of information and experiences among group members.

But it is not possible for researchers to work directly with many farmers or even in many villages in a given area; scaling up is thus often viewed as involving some tension or conflict with participation (IIRR 2000). Field practitioners in the case studies minimised this problem in several ways:

- Wambugu *et al.* (2001) worked with a range of local development partners who themselves used participatory techniques and promoted farmer experimentation and innovation. These partners included NGOs, government extension services, community-based organisations, private companies, and church organisations.

- Faminow *et al.* (2001) scaled up participatory research, helping farmers to establish 1850 test plots in 850 villages. Unique among the case studies, this project paid farmers a cash subsidy. However, the authors concluded that the high rate of uptake by farmers not receiving subsidies indicated that this incentive may not have been necessary.

- In Southern Africa, researchers helped farmers establish hundreds of farmer-designed trials, in which farmers tested new practices and species on their own and as they wished. Researchers facilitated farmer-to-farmer learning tours and monitored small samples of farmers (Böhringer 2001).

In conducting participatory extension, the case studies highlight the need for pluralistic, integrated, and bottom-up approaches (Anyonge *et al.* 2001). Wambugu *et al.* (2001) started by ensuring farmers' interest in available practices and the appropriateness of the practices to their circumstances, both biophysical and socio-economic. Böhringer (2001) noted the need to support a minimum number of farmers in an area, about 10 per cent, to catalyse uptake. Nearly all of the case studies focused on working with farmer groups rather than individuals, to economise on scarce facilitation resources and ensure greater farmer-to-farmer dissemination and exchange of information. An eclectic approach concerning extension methods was also advocated; Anyonge *et al.* (2001) found that working through schools was the most effective approach in some areas while in others, working through farmer groups was more effective.

There was also considerable variation in the degree to which practitioners in the case studies were able to involve women and also focus on the poor. In Southern Africa, Böhringer (2001) noted that facilitators with the International Centre for Research in Agroforestry (ICRAF) encouraged partners to ensure that 50 per cent of beneficiaries were women. In establishing fodder trees in central Kenya, 60 per cent of participating farmers were women (Wambugu *et al.* 2001). At the other extreme, in India, Faminow *et al.* (2001) experienced difficulty in involving women because they were excluded from owning land, and thus they were allocated only 5 per cent of the test plots. Nevertheless, the project was able to reach women by offering smaller-scale tree nurseries more suited to their needs and resources. Concerning wealth, only Noordin *et al.* (2001) in western Kenya compared the uptake of technology options among different wealth groups. While wealth was positively related to the use of fertiliser, it was not related to the use of improved fallows and biomass transfer. Because agroforestry practices require little, if any, cash outlay, they are especially suitable for resource-poor farmers.

Building local capacity

One of the most exciting achievements in the case studies has been the building of local institutional capacity, not just for implementing agroforestry but also for planning, implementing and evaluating a broad range of development activities. In local-level planning in Nyandarua and Nakuru Districts in Kenya, communities developed action plans based on their needs and designed and implemented activities together with extension staff. Many critical lessons were learned; for example, community planning was more effective through village elders and leaders of organised groups than through open public meetings. Planning exercises must take place before government staff submit their work plans so that the staff are able to commit their time to new activities (Anyonge *et al.* 2001).

In western Kenya, agroforestry researchers and development staff helped representatives of farmer groups to form village committees in order to promote the testing of practices to improve soil fertility. They also planned soil conservation activities, exchange tours with other villages, and the collective purchase of inputs. Building on existing farmer groups rather than creating new, competing structures was found to enhance impact and give the groups a sense of ownership of the process. Village committees federated into sub-location and location-level committees, and some were assisted in developing proposals, which were successful in obtaining funds for scaling-up activities. But there were also important problems. Higher-level committees were generally weaker than the village committees. Moreover, the performance of the committees was dependent on follow-up from project staff, even three years after they were formed (Noordin *et al.* 2001).

In Uganda, participatory research tools were useful in building local capacity. Agroforestry researchers and development practitioners helped communities to conduct participatory mapping exercises to plan the planting of contour hedges on hillsides to curb soil erosion and provide fodder, stakes, and fuelwood. Farmers used the maps to calculate the numbers of seedlings they needed and the numbers of seasons it would take to plant the required seedlings. They then used the information to decide how many group nurseries they needed to supply the seedlings. Such participatory methods greatly increase farmers' motivation, willingness to participate in collective action, and sense of ownership over the development process (Raussen *et al.* 2001).

In Mindanao, Philippines, farmers joined together to form Landcare groups, to share knowledge and learn more about sustainable and profitable agricultural practices that conserve natural resources. Conservation teams, made up of a farmer, an extension technician, and an outside facilitator, trained farmers and facilitated exchanges of knowledge and experiences in conservation farming practices and organisational methods. Landcare members increased rapidly in number and chapters formed associations, which sought and received funding from local governments. Their activities included establishing nurseries, training, and making farmer-to-farmer visits. Mercado *et al.* (2001) note that the greatest success of Landcare was the change in attitude of farmers and policy makers about land use and environmental protection. A second key achievement was the increased capacity of farmers to plan and implement development projects and to lobby local governments for funding and for promoting effective natural resource management. A key question for the future is, as research and extension services in many countries decline, can such farmer federations conduct their own research and development projects and manage to deliver essential services?

Germplasm

Quality planting material is needed to start scaling up, and local systems of producing and distributing planting material are needed to sustain agroforestry development. Weber *et al.* (2001) focus on the need for high-quality, genetically diverse, and appropriate planting material and describe participatory methods for developing such material. They also explain how conservation of genetic resources can take place through the use of productive, adapted, and genetically diverse planting material.

Several innovative systems of community-based seed supply and distribution are described in the case studies. In Peru, farmers are forming networks to produce and sell high-quality seed and seedlings to tree-planting projects and to timber companies (Weber *et al.* 2001). In central Kenya, facilitators are promoting community-based seed production and marketing through a range of partners: individual farmers, private nurseries, farmer groups, and seed vendors (Wambugu *et al.* 2001). Böhringer (2001) reports that ICRAF and its partners in four countries of Southern Africa are helping farmers establish 800 seed multiplication plots and 6000 nurseries in 2001.

A key and often controversial issue in scaling up is whether facilitating organisations should distribute free seed and seedlings. In most situations, small-scale subsistence farmers do not have the cash resources to pay the full cost of seed and other planting material. Yet the supply of free seed is not sustainable on a large scale, and it stifles the private nurseries that sell planting material. In Eastern Province, Zambia, organisations promoting improved fallows arrived at a viable compromise: they supply farmers with seed on condition that the farmers return twice as much seed to the organisation when it becomes available from the trees they plant, which is usually during the second year after tree establishment.

Market options

Among the ten elements of scaling up, the case studies are probably weakest in developing market options. Most do not even mention the role of markets. Many of the agroforestry practices assessed in the case studies do not yield products for sale; rather, they provide substitutes for purchased inputs, such as fodder shrubs for dairy feeds or improved fallows for mineral fertiliser. Thus, issues concerning product markets are not directly related to their promotion and development. But other agroforestry products such as fruit and timber may be sold, and the potential benefits from transforming and marketing them are often huge. Böhringer (2001) noted that researchers in Southern Africa are beginning to assess market demand and consumer preferences for indigenous fruits, so that mechanisms can be put in place for establishing links between producers and markets. Assessments are being made of selling fresh fruit as well as producing jams, juices, and alcoholic beverages.

Most agroforestry research and development teams, in fact, lack skills in marketing and product development. Gaining access to such expertise needs to be a high priority in scaling up. Lecup and Nicholson (2000) provide useful guidelines for identifying market opportunities for agroforestry products. Franzel and Denning (in press) identify key elements of successful marketing and present a conceptual framework of marketing research and development for scaling up agroforestry innovations.

Policy options

An enabling policy environment is critical for scaling up. Whereas policy research often focuses on the national level, the case studies highlighted the importance of a range of local policy makers, both traditional and governmental, in villages, districts, and provinces. These local policy makers proved to be at least as important for promoting the scaling up of agroforestry as national policy makers based in the capital city.

Agroforestry researchers and development staff in the case studies helped inform policy makers about policy constraints, which has led to the constraints being removed. For example, in parts of Kenya, ordinances require farmers to obtain a permit before cutting down trees, on the seemingly logical assumption that such measures protect trees. But they are actually a strong disincentive against planting trees, since farmers do not want to plant trees that they may not be able to harvest. Moreover, the ordinances are often abused, as farmers are required to negotiate their way through bureaucracies or even pay bribes to obtain the cutting permits. Anyonge *et al.* (2001) described how agroforestry development staff were able to persuade the provincial administration to make redundant the permits needed to cut trees and thus remove this strong disincentive for planting them.

Also of importance, the case studies demonstrate how local policy makers in particular can act to promote agroforestry. In Nagaland, India, village leaders passed resolutions supporting tree planting, which greatly influenced farmers' decisions (Faminow 2001). In Mindanao, Philippines, local governments provided funds, technical assistance, and policy support for conservation practices. Municipalities developed natural resource plans and they funded conservation teams and Landcare association activities such as nurseries, training, and cross-site visits (Mercado *et al.* 2001).

In Kabale, Uganda, local policy makers played a lead role in scaling up agroforestry. Local leaders are elected, and their re-election depends in great part on their ability to promote development activities for their constituency. The government's ambitious decentralisation programme provided considerable authority and funds to local government councils, which often had a strong interest in agroforestry as a means for improving household incomes and conserving natural resources. Project staff linked with local policy makers in numerous activities to scale up agroforestry, including planning, mobilising the community, and producing community newsletters (Raussen *et al.* 2001). Even in countries with weak local governments, great potential exists to mobilise local authorities to promote agroforestry development.

Successful local pilot projects may also be scaled up to the national level through policy contacts. For example, experience in local-level planning in a development project in Kenya played a key role in developing a national extension programme, which involved greater participation by local stakeholders in planning and budgeting local-level extension programmes throughout the country (Anyonge *et al.* 2001).

Learning from successes and failures

Monitoring and evaluation served to enhance learning among stakeholders in all of the case studies. Many examples were provided about the ways in which feedback from farmers resulted in important modifications in recommendations, strategies, and policies. Faminow *et al.* (2001) report that the low adoption rate of labour-intensive contour bunds resulted in a shift in project direction towards farmers' own measures for soil erosion, which was to use small trenches. Low adoption rates by women led to special emphasis to find out their needs and to tailor tree-planting strategies to meet those needs. The high rate of tree planting among farmers who were not involved in trials indicated the project's success. Moreover, surveys monitoring farmer plantings helped indicate farmers' preferences for trees that were the most marketable. These findings helped the project to better meet farmers' needs.

Böhringer (2001) presented the idea of pilot development projects as laboratories to understand impact under real-world conditions. In Malawi, he is assessing whether farmers can adopt agroforestry to control soil erosion and is investigating hypotheses concerning gender, wealth, researcher-to-farmer and farmer-to-farmer communication, and the role that community organisations play in promoting adoption.

The case studies also assessed the impact of scaling up. Anyonge *et al.* (2001) explained how aerial surveys in Kenya were used to show that the useable volume of wood in project areas doubled in five years. Wambugu *et al.* (2001) reported the economic benefits accruing to farmers adopting fodder trees and the huge potential benefits nationally if just half of Kenya's dairy farmers were to adopt them. Such analyses provide important arguments to planners and donors for investing further in scaling up tree planting for improving farmer incomes and livelihoods. But there was no clear evidence as to how increased income was actually spent or how it benefited the households. None of the case studies presented values for environmental impact, although several had environmentally linked objectives.

Furthermore, while the case studies emphasise project monitoring and evaluation, little attention was given to farmers' own monitoring and evaluation. Böhringer (2001) describes monitoring and evaluation by three types of actors: individual farmers, development agents, and villagers in workshops. Triangulation among these three approaches would give a more accurate picture of successes and failures than any single one of them alone. Kristjanson *et al.* (in press) describe the importance of farmer workshops for identifying farmers' expectations about the impact arising from adoption of improved practices and farmers' proposed impact indicators.

Knowledge and information sharing

Sharing knowledge and information is critical to ensure effective decision making by a wide range of stakeholders in the scaling-up process (Cooper and Denning 2000). Farmers' indigenous knowledge played an important role in shaping tree domestication in Peru. For example, farmers were adept at distinguishing among *Bactris gasipaes* (peach palm) varieties. They could associate physical attributes such as waxy coats with desired fruit characteristics, such as oil content. Such information was useful for helping researchers to select which varieties to multiply.

Even more important is for farmers to share knowledge among themselves. Böhringer (2001) reported facilitating farmer-to-farmer group training exercises, in which participants spend several days visiting farmers in another village, sharing knowledge along with board and lodging. Such training exercises cost about one-tenth as much per person trained as do formal training courses. Training and supporting farmer trainers is another key means for promoting farmer-to-farmer knowledge sharing.

The case studies also documented considerable farmer modification of introduced practices. In India, for example, farmers chose to plant timber trees more densely than recommended for several reasons: to reduce weeding, to get straighter trunks, and to reduce soil erosion (Faminow *et al.* 2001). Farmers in central Kenya found that extending the time that *Calliandra calothyrsus* seeds were soaked increased germination. This information was fed back to researchers, who confirmed the validity of the finding. Extensionists now recommend the longer soaking time (Wambugu *et al.* 2001). Continuous farmer experimentation, adaptation, and knowledge sharing are critical to ensure that practices are appropriate over large areas (Böhringer 2001).

Strategic partnerships and facilitation

Most of the case studies put great emphasis on partnerships as a means for scaling up. Most also are written from the point of view of a facilitator assisting a range of partners. Böhringer (2001) noted that ICRAF collaborates with 572 partner organisations in four Southern Africa countries in scaling up agroforestry practices. But he also pointed out that numbers are not what is important; rather, detailed analyses are needed to assess the quality of partnerships, that is, what have been the successes and the failures, and how can high transaction costs be reduced. Noordin et al. (2001) cites several challenges in building partnerships: drawing up clear memoranda of understanding on roles and responsibilities, reducing duplication of effort, reducing partners' expectations about the material benefits they will receive through collaboration, and improving the documentation of activities.

Table 1 is an example of a matrix, adapted from Tanzania, for helping facilitators assess the potential contribution of different partners in agroforestry dissemination. The matrix helps to characterise partners, to compare their strengths and weaknesses, and to decide systematically how much effort to give to each.

Böhringer (2001) highlights the special case of government partners, such as extension services, which are often weak and have top-down approaches in working with farmers. Collaborating with extension staff often requires paying substantial staff allowances to compensate for their low salaries. Yet it is often politically necessary to work closely with them. He suggests that extension roles be redefined to facilitate and coordinate services rather than to deliver them.

Other case studies report on more effective partnerships with governments, especially local government. As mentioned in the section on policy options, agroforestry researchers and development staff have built effective partnerships with local authorities in Uganda and the Philippines. In Mexico, Haggar *et al.* (2001) reported working effectively with government development projects that were training extensionists in participatory methods and agroforestry practices.

Table 1: Matrix for assessing the potential contribution of different partner organisations in agroforestry dissemination (partners can be scored high, medium, or low on each criterion)

	Partner organisation		
	1	**2**	**3**
Reach (areas and no. of farmers)	H	L	H
Participatory approaches	L	H	H
Availability of staff, resources, good management	H	M	M
Commitment to agroforestry	H	H	L
Openness to appropriate practices	L	H	M
Commitment to monitoring and evaluation	M	M	L
Accessibility (distance)	H	M	L
Shared objectives	H	H	M
Time and resources that ICRAF spends on them	L	H	L
Value per unit effort	High, if participatory approach can be introduced	Low, partner is very small	Partner has many activities and is not very interested in agroforestry

A main instrument for facilitating partnerships in Southern Africa has been 'networkshops' – informal, biannual meetings at which representatives from partner organisations and farmers plan and review their agroforestry activities (Böhringer 2001). A most important impact of the workshops is that all partners develop a sense of involvement, enthusiasm, and ownership of promising innovations. A critical task of the networkshops is to define clearly the roles and responsibilities of the different actors in on-farm research and dissemination. In Eastern Province, Zambia, about 75 representatives of research, extension, NGOs, and farmer groups have met once or twice a year since 1996 in networkshops to plan and review the testing and dissemination of improved fallows and other agroforestry practices. Five different organisations provide funding, and network-shops are hosted on a rotational basis and chaired by the provincial coordinator of agriculture.

Whereas in 1996 the ICRAF-Zambia project was seen as the main facilitator of agroforestry in Eastern Province, it has helped build capacity in several other organisations, which now provide seed, training, and technical assistance. The project has evolved from being the hub of agroforestry activity to becoming one of several nodes of the network. This evolution attests to its successful role as a facilitator. Noordin *et al.* (2001) report a similar effort launched in 2001, called the Consortium for Increasing Farm Productivity in Western Kenya. Planned activities include scaling up improved practices, sharing methods and approaches, developing training materials, and issuing a newsletter.

Research challenges on scaling up

The case studies presented in this issue demonstrate the multifaceted nature of scaling up: temporal, spatial, institutional, and functional facets, to name just a few. A key lesson is that scaling up agroforestry is not just transferring inputs and knowledge about improved practices; it involves building partnerships, assisting communities to mobilise resources, and promoting effective participation of stakeholders to test, disseminate, adapt, and evaluate new innovations in a sustainable manner (Wambugu *et al.* 2001). Böhringer (2001) draws a similar conclusion: in addition to offering improved livelihoods, agroforestry is a learning tool for building local capacity to innovate.

Review of the case studies reveals several challenges ahead for enhancing the scaling up of agroforestry. An overarching problem is that there is a paucity of research on the scaling-up process. Whereas many useful lessons can be derived from the cases presented here, they are almost always based on informal analysis – the reflections of practitioners – rather than on rigorously planned research. Yet careful assessments of the relative costs and benefits and the advantages and disadvantages of different strategies are often possible. Resources dedicated to project or programme monitoring and evaluation could be used or supplemented to investigate the effectiveness of scaling-up processes, and not just the inputs and outputs. In addition, wherever possible, opportunities should be taken to undertake simple planned comparisons of different approaches. Based on the conclusions of these case studies and evidence from the broader literature, the following issues need to be addressed as a matter of priority:

- Scaling up requires a continuous stream of technical options based on both science and farmer innovation. How do we capture farmer innovation and ensure that scientific knowledge and indigenous knowledge are well integrated?

- In the process of scaling up, farmers adapt and improve innovations as they are extended to different circumstances and face different resource constraints and stresses. How can monitoring and evaluation systems be designed to capture the knowledge generated in this way?

- Which information dissemination methods are most effective and why? For example, how do the costs and benefits of farmer-to-farmer visits compare with those of farmer training courses?

- Many different models for empowering local communities as change agents were presented in the case studies. What are the guiding principles for successful and sustainable farmer organisations? How can we help such organisations to federate across villages to improve their efficiency and effectiveness?

- What are the advantages and disadvantages of different means of producing and distributing or marketing seed at different stages of the scaling-up process? How can community-based production and marketing of seed be made institutionally and financially sustainable?

- Marketing agroforestry products is an untapped strategy. How can we link farmer production to local, regional, and international markets?

- How can policy makers – at various levels – become effective promoters of local farmer organisations and agroforestry development? The strategy for involvement will depend on the level at which the policy maker is operating.

- What is the impact of agroforestry practices on the livelihoods of women and poor households and on the environment? How can we facilitate farmer and community-based monitoring and evaluation?

- How can we devise more strategic partnerships and reduce their transaction costs? How can issues of institutional ownership and attribution be overcome for the benefit of small-scale farmers?

- How can research institutions adapt functionally and structurally to be more effective partners in scaling up and, more broadly, in rural development?

Uvin and Miller (1999) claim that 'scaling up' is akin to the Loch Ness monster – many have sighted it, but its description is as varied as the people who have written about it. Unfortunately, but not surprisingly, there is no definitive formula for scaling up. Yet this analysis of case studies, when considered in conjunction with the earlier syntheses by Cooper and Denning (2000) and IIRR (2000), demonstrates a convergence on the elements that, to various degrees, are important in the process. Regional, national, and local specificities clearly suggest that greater investment is warranted in the learning and innovation associated with scaling up.

References

Anyonge, T.M., Christine Holding, K. K. Kareko, and J. W. Kimani (2001) 'Scaling up participatory agroforestry extension in Kenya: from pilot projects to extension policy', *Development in Practice* 11(3–4): 449-59, and this volume

Böhringer, Andreas (2001) 'Facilitating the wider use of agroforestry for development in Southern Africa', *Development in Practice* 11(3–4): 434-48, and this volume

Cooper, Peter and Glenn L. Denning (eds) (2000) *Scaling Up the Impact of Agroforestry Research*, report of the Agroforestry Dissemination Workshop, Nairobi, Kenya, 14–15 September 1999, Nairobi: International Centre for Research in Agroforestry

Denning, Glenn L. (2001) 'Realising the potential of agroforestry: integrating research and development to achieve greater impact', *Development in Practice* 11(3–4): 407-16, and this volume

Faminow, Merle D., K.K. Klein, and Project Operations Unit (2001) 'On-farm testing and dissemination of agroforestry among slash-and-burn farmers in Nagaland, India', *Development in Practice* 11(3–4): 471-86, and this volume

Franzel, Steven and Glenn L. Denning (eds) (in press) *Key Elements of Successful Marketing Research and Development for Scaling Up Agroforestry Innovations*, proceedings of an international workshop held at ICRAF, Nairobi, Kenya, 16–17 September 1999, Nairobi: International Centre for Research in Agroforestry

Haggar, Jeremy, Alejandro Ayala, Blanca Díaz, and Carlos Uc Reyes (2001) 'Participatory design of agroforestry systems: developing farmer participatory research methods in Mexico', *Development in Practice* 11(3–4): 417-24, and this volume

International Institute of Rural Reconstruction (IIRR) (2000) *Going to Scale: Can We Bring More Benefits to More People More Quickly?*, Silang, Cavite, Philippines: IIRR

Krishna, A., N. Uphoff, and M.J. Esman (eds) (1998) *Reasons for Hope: Instructive Experiences in Rural Development*, Bloomfield CT: Kumarian

Kristjanson, Patricia, Frank Place, Steven Franzel, and P.K. Thornton (in press) 'Assessing research impact on poverty: the importance of farmers' perspectives', *Agricultural Systems*

Lecup, I. and K. Nicholson (2000) *Community-Based Tree and Forest Product Enterprises: Market Analysis and Development*, Rome: FAO

Mercado, Agustin R., Jr., Marcelino Patindol, and Dennis P. Garrity (2001) 'The Landcare experience in the Philippines: technical and institutional innovations for conservation farming', *Development in Practice* 11(3–4): 495-508, and this volume

Noordin, Qureish, Amadou Niang, Bashir Jama, and Mary Nyasimi (2001) 'Scaling up adoption and impact of agroforestry technologies: experiences from western Kenya', *Development in Practice* 11(3–4): 509-23, and this volume

Quiñones, M. and T. Gebre (1996) 'An overview of the Sasakawa–Global 2000 Project in Ethiopia', in S.A. Breth (ed.) *Achieving Greater Impact from Research Impacts in Africa*, Mexico City: Sasakawa Africa Association

Raussen, Thomas, Geoffrey Ebong, and Jimmy Musiime (2001) 'More effective natural resource management through democratically elected, decentralised government structures in Uganda', *Development in Practice* 11(3–4): 460-70, and this volume

Uvin, P. and D. Miller (1999) 'Scaling up: thinking through the issues', available at http://www.brown.edu/Departments/World_Hunger_Program/hungerweb/WHP/SCALINGU.html

Wambugu, Charles, Steven Franzel, Paul Tuwei, and George Karanja (2001) 'Scaling up the use of fodder shrubs in central Kenya', *Development in Practice* 11(3–4): 487-94, and this volume

Weber, John C., Carmen Sotelo Montes, Héctor Vidaurre, Ian K. Dawson, and Anthony J. Simons (2001) 'Participatory domestication of agroforestry trees: an example from the Peruvian Amazon', *Development in Practice* 11(3–4): 426-33, and this volume

Resources

While the contributors to this volume focus on the subject area of agroforestry, they do so from a perspective that places people – their communities, their forms of organisation, their livelihoods, and their knowledge systems – at the centre. So this book deals as much with how to facilitate and scale up the impact of participatory or farmer-led research in ways that make a real difference to the lives of those involved as it does with agroforestry per se. In compiling this annotated resources list, we have sought to reflect this same perspective, favouring works that explore issues and approaches relating to participatory or farmer-led research over material of a more technical nature. For the benefit of readers who are interested in agroforestry as such, we have also included entries on a number of specialised organisations that publish practical resources, such as manuals and handbooks, or undertake academic research.

The list was compiled and annotated by Nicola Frost, with Deborah Eade and Alina Rocha Menocal, all editorial staff at Development in Practice, *with additional input from Steven Franzel, co-editor of this volume and based at ICRAF.*

Books

Carine Alders, Bertus Haverkort, and Laurens van Veldhuizen (eds.): *Linking with Farmers: Networking for Low-External-Input and Sustainable Agriculture,* London: Intermediate Technology, 1993.

Offering a perspective on farmer-led extension, this book gives examples of a range of networking activities undertaken by farmers' groups across the world. Enhanced farmer-to-farmer exchange can be an excellent way of promoting low-external-input extension, and although the book also acknowledges the organisational difficulties that can arise, it argues that a shift in emphasis away from external input opens up new approaches to agricultural research.

J. E. Michael Arnold and Peter A. Dewees (eds.): *Farms, Trees and Farmers: Responses to Agricultural Intensification*, London: Earthscan, 1997.

This book sets out to examine the role of trees grown on farms in developing countries, particularly in light of the increasing intensification of agriculture. The two central sections survey trends in tree-growing by farmers and consider the factors which influence their decision-making. Case studies from Kenya and South Asia cover, among other topics, the need for tree products and the nature of incentives to grow trees, the importance of adequate market access, the allocation of land and labour within the household, and exposure to risk. (Previously published as *Tree Management in Farmer Strategies*, Oxford: OUP, 1995.)

Solon L. Barraclough and Krishna B. Ghimire: *Forests and Livelihoods: The Social Dynamics of Deforestation in Developing Countries*, Basingstoke: Macmillan Press and New York: St Martin's Press, 1995 (in association with UNRISD).

Based on research and detailed case studies in Brazil, Central America, Nepal, and Tanzania, this book argues that conventional or single-factor explanations for increased rates of deforestation (such as population growth, ignorance on the part of peasant farmers, or market and policy failures) are over-simplistic. The success of technical solutions depends on incorporating these into policies to manage forest areas and natural resources in order to meet social goals on a more equitable basis. See also *Agricultural Expansion and Tropical Deforestation: Poverty, International Trade and Land Use*, London: WWF and Earthscan, 2000.

William R. Bentley, P. K. Khosla, and Karen Seckler: *Agroforestry in South Asia: Problems and Applied Research Perspectives*, New Delhi: Oxford & IBH Publishing, 1993.

Part I of this book considers the physical and biological aspects of agroforestry in South Asia. Part II analyses the ways in which structural constraints, including cultural, economic, and social variables, affect – and sometimes even override – objectively sound technical programmes.

C. den Biggelaar: *Farmer Experimentation and Innovation: A Case Study of Knowledge Generation Processes in Agroforestry Systems in Rwanda*, Community Forestry Case Study 12, Rome: FAO, 1996.

Although farmers in Rwanda have always used trees for numerous purposes, the active planting and management of woody vegetation on farms is relatively recent. This case study seeks to explore and understand farmers' processes of generating knowledge of agroforestry that underlie these changes in resource management and use, with particular focus on experimental methods used by farmers in integrating trees into their farms. This study is the first in a series on farmer-initiated research and experimentation, the goal of which is to identify more effective ways in which farmers can be supported in their own experimentation and knowledge-sharing, while also working towards a consolidation of local forestry knowledge.

Jean-Marc Boffa: *Agroforestry Parklands in Sub-Saharan Africa*, FAO Conservation Guide, No. 34, Rome: FAO, 1999.

This publication provides a broad overview of agroforestry parkland systems, analysing their operation in a wide variety of geographical settings and assessing their social impact and economic significance at both the local and national levels.

Louise E. Buck, James P. Lassoie, and Erick C. M. Fernandes (eds.): *Agroforestry in Sustainable Agricultural Systems*, Boca Raton FL: CRC Press, 1999.

Concentrating on successful strategies for raising forests and tree products for commercial harvesting and land and watershed management, this book examines the environmental and social conditions necessary for sustainable agroforestry. It analyses a wide variety of ecological settings in great detail, including, for example, the combination of agroforestry with livestock systems. The book may also be useful as a textbook for students.

P.J.M. Cooper and G.L. Denning: *Scaling Up the Impact of Agroforestry Research*, Nairobi: ICRAF, 2000.

New agroforestry practices, combining scientific and indigenous technical knowledge, have made important contributions to improving household welfare and land use in many areas. This publication addresses the issue of how to spread the benefits of agroforestry to larger numbers of people, in a sustainable and equitable manner. Ten key elements of scaling up are described: technical options, farmer-centred research and extension, local institutional capacity, germplasm, market and enterprise development, policy options, learning from successes and failures, strategic partnerships, knowledge sharing, and facilitation.

Craig R. Elevitch and Kim M. Wilkinson (eds.): *The Overstory Book: Cultivating Connections with Trees*, Holualoa, Hawaii: Permanent Agriculture Resources, 2001.

This collection of the first three years of the electronic journal *The Overstory* is fully indexed, organised by topic, and cross-referenced, and it also benefits from a resources section and an index of botanical names. It forms a practical manual for agroforestry techniques and combines technical sections on soil, seed selection, and livestock with sections on valuing indigenous knowledge, marketing, and contributions to human health. Contributors include Peter Huxley, P. K. Ramachandran Nair, Helen van Houten, and Roland Bunch. The journal can be viewed online at www.agroforester.com.

Cornelia Flora (ed.): *Interactions between Agroecosystems and Rural Communities*, Boca Raton FL: CRC Press, 2001.

There is an increasing realisation among biophysical scientists that human behaviour critically influences the extent to which agro-ecosystems are implemented. This book examines this relationship and offers an understanding of alternative ways of working with communities to increase agro-ecosystem sustainability. Through a general overview and a series of case studies, the book explores the way in which changes in the local economy can affect support for agricultural innovation. It also addresses specific community-based actions in both temperate and tropical zones in Europe,

North America, Asia, Central America, and Latin America that have resulted in more sustainable agro-ecosystems.

Steven Franzel and Sara J. Scherr (eds.): *Trees on the Farm: Assessing the Adoption Potential of Agroforestry Practices in Africa*, Wallingford: CABI Publishing, 2002.

Following rigorous scientific methods, this edited volume sets out to analyse agroforestry in an innovative manner by focusing not only on its biophysical aspects, as other studies have done, but also on its socio-economic dimensions. Drawing on select methodologies and participatory field research from five case studies conducted in Kenya and Zambia, the book analyses the adoption potential of promising agroforestry practices in Africa and highlights the importance of policies for enhancing adoption. It also presents and explains methods that researchers and field practitioners could use to assess adoption potential and draws lessons for improving the effectiveness and efficiency of developing and disseminating agroforestry technologies.

Andrew M. Gordon and Steven M. Newman (eds.): *Temperate Agroforestry Systems*, Wallingford: CABI, 1997.

While much of the recent research on agroforestry has been carried out in the tropics and within the context of developing nations, this book explores the development of agroforestry in temperate zones, analysing in particular the role of agroforestry in silvopastoral and cropping systems and in the promotion of soil conservation. Case studies concentrate on those areas where the greatest advances, adoptions and modifications have taken place, namely the Americas, China, Australasia, and Europe.

Karin Hochegger: *Farming Like the Forest: Traditional Home Garden Systems in Sri Lanka*, Weikersheim: Margraf Verlag, 1998.

Forest garden systems in South and South-East Asia have long proven to be a highly productive and sustainable form of agriculture, but, despite their proven efficiency over the centuries, research on such models has been minimal. However, growing concern about the destruction of tropical ecosystems has led planners and agricultural scientists to turn their attention to traditional practices in search of solutions to contemporary problems. This book describes the system in Sri Lanka, where forest gardens have contributed to a balanced and harmonious relationship between people and nature, with the aim of informing future agricultural models elsewhere.

Edvard Hviding and Tim Bayliss-Smith: *Islands of Rainforest: Agroforestry, Logging and Eco-tourism in the Solomon Islands*, Aldershot: Ashgate, 2000.

This book offers an ethnographic and historical study of land use in the Solomon Islands which provides a rich account of the complexity and longevity of indigenous agroforestry systems. The second half, which examines how global trends may affect local-level operations, is especially valuable. In this section, for example, the authors discuss how the arrival of Asian logging companies turned the forest into a commodity and led to concomitant political and environmental problems. The study serves as a reminder that agroforestry models do not operate in a vacuum, but rather evolve within structural contexts in which real-world situations must be reckoned with.

Charles V. Kidd and David Pimentel (eds.): *Integrated Resource Management: Agroforestry for Development*, San Diego CA: Academic Press, 1992.

As its starting point, this book argues that the techniques of the Green Revolution are not applicable to large parts of the world, and that there is an urgent need to improve the food security of the growing number of people dependent on small areas of marginal land. The contributors suggest that agroforestry techniques can help to stabilise incomes and levels of food production without high levels of external input. The book not only undertakes a detailed cost/benefit analysis of agroforestry and addresses technical issues of soil and water resource management, but also considers some of the social and cultural issues at stake.

H.J.W. Mutsaers, G.K. Weber, P.Walker, and N.M. Fisher: *A Field Guide for On-Farm Experimentation*, The Hague: International Service for National Agricultural Research, 1997.

Applied agricultural research has conventionally been carried out in specialist research stations, while development organisations were expected to transform the results of this 'lab-based' research into practical solutions for farmers. However, it has by now been recognised that this technique does not take account of the many constraints, both physical and socio-economic, within which poor farmers operate. This manual provides a practical guide to the operation of successful on-farm research as an essential tool in the development and transfer of agricultural innovation.

P.K. Ramachandran Nair and C.R. Latt (eds.): *Directions in Tropical Agroforestry Research*, Dordrecht: Kluwer, 1998.

This volume is a compilation of ten reviews of tropical agroforestry research projects carried out over the past two decades. Each paper synthesises the results of research, summarises the current state of knowledge, identifies knowledge gaps, and outlines directions for future research. Examples come from Brazil, sub-Saharan Africa, Asia, and the Pacific.

Gordon Prain, Sam Fujisaka, and Michael D. Warren (eds.): *Biological and Cultural Diversity: The Role of Indigenous Agricultural Experimentation in Development*, London: Intermediate Technology, 1999.

Contributors to this edited volume illustrate the intimate relationship between biological and cultural diversity, with case studies demonstrating the range of farmer experimentation, the depth of knowledge that farmers possess about their local environment, and the importance of recognising the ultra-local, site-specific nature of much of this knowledge and innovation, an issue often overlooked because of constant pressure for successful innovations to 'scale up'.

Richard K. Reed: *Prophets of Agroforestry: Guarani Communities and Commercial Gathering*, Austin TX: University of Texas, 1995.

In this book, the author argues that the economic and social basis for the relative autonomy of the Chiripá (Guaraní) people of eastern Paraguay lies in commercial agroforestry. Resisting the pressure to clear land for commercial agriculture, the Chiripá harvest and sell forest products without destroying the forests. Reed also explores the ways in which Chiripá social organisation, which centres on kin ties, facilitates the necessary adaptation to the challenges and opportunities posed by the commercialisation of agriculture. Although departing slightly from the themes of on-farm agroforestry, this study provides an alternative model for the sustainable management of subtropical forests.

N.C. Saxena and Vishwa Ballabh (eds.): *Farm Forestry in South Asia*, New Delhi: Sage India, 1995.

This volume concentrates on the perspectives and decision-making processes of farmers in South Asia in relation to farm forestry. Analysing the nature of indigenous agroforestry initiatives, the book suggests a model for future projects that combines a clear sense of market priorities with adequate provision for subsistence needs. Advocating the reversal of previous recommendations, the contributors call for the use of private land for market-orientated short-rotation products, while reserving public land for fodder, fuelwood, and consumption. Case studies examine the socio-economic implications of these observations. See also N.C. Saxena and Naresh Chandra, *Forests, People and Profit*, 1995.

Vanessa Scarborough, Scott Kilough, Debra A. Johnson, and John Farrington (eds.): *Farmer-Led Extension: Concepts and Practices*, London: ITDG Publishing, 1997.

A stimulating combination of case study and analysis, this book provides an excellent introduction to the principles and practice of farmer-led extension. A discussion of current challenges to agricultural extension provides a useful context against which to examine the differing approaches in South and South-East Asia and Latin America. The book also considers the role of NGOs and governments in supporting farmer-led initiatives, and it assesses the constraints and possibilities for scaling up and expanding networks.

Ralph Schmidt, Joyce K. Berry, and John C. Gordon: *Forests to Fight Poverty: Creating National Strategies*, New Haven CT: Yale University Press, 1999.

Rather than focusing on agroforestry as such, this book concentrates on policy strategies to combat the problem of deforestation, especially in tropical forests. The authors provide a useful insight, from a policy perspective, into the links between land access, the quality of forest land, and poverty. Arguing that trees can contribute to secure livelihoods, the book makes a strong case for seeing trees and forests as important potential tools for poverty alleviation. The authors also call for an inclusive, participatory approach to strategic planning in agroforestry that recognises the importance of country-specific plans.

Richard A. Schroeder: *Shady Practices: Agroforestry and Gender Politics in The Gambia*, Berkeley CA: University of California Press, 1999.

The result of detailed long-term research in Gambia, this book explores the complex gender relations exposed by the conflict between two ostensibly 'progressive' development programmes. One programme is designed to promote market gardening as a livelihood strategy for women, while the other calls for the introduction of agroforestry practices by men in low-lying areas. Eventually, however, the latter came to threaten the market gardens, once the trees grew and the shade canopy closed. Women therefore saw agroforestry as a means for men to reclaim control over the land and undermine the social gains made by the women through gardening. The book provides an intricate case study of the need to consider local social relations in any planned extension work and of the changing emphasis on gender in development.

Ian Scoones and John Thompson (eds.): *Beyond Farmer First: Rural People's Knowledge, Agricultural Research and Extension Practice*, London: ITDG Publishing, 1994.

As its title suggests, this book takes forward many of the principles established in *Farmer First* (1989), edited by Robert Chambers *et al*. Through a series of brief papers from a large number of contributors, the current volume undertakes a more detailed analysis of how differences based on gender, age, class, and other social categories affect access to and control of natural resources. The book goes on to ask what the institutional and policy implications of privileging farmer-led agricultural research may be.

M.P. Singh and D.N. Tewari: *Agroforestry and Wastelands*, New Delhi: Anmol Publications, 1996.

Increasing pressure on land resources and rapid deforestation in India have led to a sharp increase in areas classified as wastelands. Animal husbandry, agriculture, and forestry have all developed separately, despite their common relation to land – and they often operate in conflict with one another. Very often, it is the trees that suffer, despite their critical importance to rural livelihoods. The authors of this book call for re-connecting these elements, both at the village and policy levels, and combining local knowledge and experience with scientific modelling and an improved genetic base to regenerate degraded land and develop sustainable land-management systems. See also D.N. Tewari: *Agroforestry for Increased Productivity, Sustainability and Poverty Alleviation*, Dehradun, India: International Book Distributors, 1995.

Panjab Singh, P. S. Pathak, and M. M. Roy (eds.): *Agroforestry Systems for Sustainable Land Use*, New Delhi: Oxford & IBH Publishing, 1994.

Bringing together papers from a conference on agroforestry and degraded lands, this collection of essays provides case studies from the Asia-Pacific region, while including some examples from Europe. Contributors investigate ways in which agroforestry can be useful in providing alternatives to unsustainable slash-and-burn (swidden) agriculture, in regenerating degraded slopes and grasslands, and even in mitigating climate change. Anil Gupta's 'Ten Myths About Agroforestry' provide an accessible introduction to the most prevalent misconceptions and objections.

James Sumberg and Christine Okali: *Farmers' Experiments: Creating Local Knowledge*, Boulder CO: Lynne Rienner, 1997.

A contribution to the debate about indigenous knowledge and farmer experimentation in agricultural development, this book aims to characterise and set in context farmers' experimentation in Africa, in order to contribute to an empirical and theoretical base from which to evaluate alternative models for the interaction between formal research and farmer innovation. The authors attempt to move beyond the general issues concerning farmer participation, which have been much discussed, to a more detailed analysis of a combined approach promoting synergy between farmers and researchers.

Laurens van Veldhuizen, Ann Waters-Bayer, and H. de Zeeuw: *Developing Technologies with Farmers: A Trainer's Guide for Participatory Learning*, London: Zed Books, 1997.

This manual is intended for NGO and government agency trainers who are preparing their staff to work together with farmers in developing technologies appropriate to ecological agriculture, while relying on few external inputs. The training is designed to stimulate interactive learning among participants, based on their own experiences.

Laurens van Veldhuizen, Ann Waters-Bayer, Ricardo Ramírez, Debra A. Johnson, and John Thompson (eds.): *Farmers' Research in Practice: Lessons from the Field*, London: ITDG Publishing, 1997.

This edited volume is a valuable compendium of practical case studies, showing farmer-led research in action in a variety of contexts, with examples examining the nature of on-farm innovation, some instances of external support for experimentation, experience of refining experimental design, and encouraging sustainability. Some of the challenges identified include the questions of how to scale up these approaches and how to influence policy content and the decision-making process.

Paul A. Wojtkowski: *The Theory and Practice of Agroforestry Design: a Comprehensive Study of the Theories, Concepts and Conventions that Underlie the Successful Use of Agroforestry*, Enfield NH: Science Publishers, 1998.

An advanced text with an emphasis on theoretical issues, this book presents a detailed look at the concepts, principles, and practices that underlie the application of agroforestry systems. The focus is on how the individual theories and concepts contribute to the process of designing or understanding user-specific agroforestry systems, and how theory influences or leads to successful application.

Journals

Agroforestry Abstracts: published monthly on the internet by CABI, in association with ICRAF.

Approximately 1500 abstracts on all aspects of agroforestry, from discussions of trees and crops to research methods and socio-cultural dimensions, are added every year.

Agroforestry News: published quarterly by the Agroforestry Research Trust, ISSN: 0967-649X.

Articles provide detailed information about cultivating a number of trees and shrubs in a temperate climate. The Trust's website includes some introductory information on temperate agroforestry systems. www.agroforestry.co.uk

Agroforestry Systems (incorporating Agroforestry Forum): published nine times per year by Kluwer in co-operation with ICRAF, ISSN: 0167-4366. Editor: P.K. Ramachandran Nair.

Publishes results of original research and critical reviews on both biophysical and socio-economic aspects of agroforestry. Recent issues have contained papers on research methodologies and techniques, including on-farm evaluation and farmer assessment.

Agroforestry Today: published quarterly by ICRAF. ISSN: 1013-95910255-8173.

Carries reports from around the world on trees and crops on farms, and on the people who plant them. Articles assess new agroforestry technologies that researchers are developing together with farmers, and indigenous agroforestry systems that farmers themselves are using successfully. The periodical also aims to highlight new research findings and assess their potential benefits for farmers. A Chinese-language version is published by the Institute of Soil Science, Academia Sinica, China.

Inside Agroforestry: published quarterly by the National Agroforestry Research Center (NAC).

A US-focused newsletter with brief articles on many aspects of agroforestry practices. Recent issues have considered the use of trees as carbon sinks, to protect watersheds, and to prevent erosion. Several titles are available in Spanish. *Agroforestry Notes*, a series of technical notes, is also published by the NAC. Both are available free online at www.unl.edu/nac

Organisations

Amazon Agroforestry Research Centre (Centro de Pesquisa Agroflorestal da Amazônia Ocidental, CPAA): Founded in 1989 as an institutional branch of the Brazilian Agricultural Research Corporation (Empresa Brasileira de Pesquisa Agropecuária, EMBREPA), CPAA seeks to generate and disseminate scientific and technical information to support the sustainable development of the Amazon through rational land use and the conservation of renewable natural resources. While most of its work is in Portuguese, the Centre is currently translating its webpage to make its information available in English. Contact details: Rodovia Am – 010, km 29, C.P. 319, Cep 69.011.970 – Manaus, Amazonas, Brazil. E-mail: sac@cpaa.embrapa.br; Web: www.cpaa.embrapa.br

Asia-Pacific Agroforestry Network (APAN): A network for the exchange of information on agroforestry research, development, and training in the Asia-Pacific region, with special reference to the 11 participating countries, APAN also co-ordinates regional and national agroforestry training courses for trainers, extensionists, and researchers. Contact details: FAO–APAN, PO Box 48, Bogor 16004, Indonesia. E-mail: fao-apan@cgiar.org; Web: www.apan.net

Association for Strengthening Agricultural Research in Eastern and Central Africa (ASARECA): A network of ten national agricultural research institutes, established in 1993 to improve regional collaboration and facilitate more efficient use of research resources. Its quarterly newsletter, *Agriforum*, is available online. The Association also operates through a variety of research networks, including the Trees-On-farm Network (TOFNET), which specialises in agroforestry. Contact details: ASARECA: PO Box 765, Entebbe, Uganda. E-mail: asareca@imul.com; Web: www.asareca.org. TOFNET: Daniel Nyamai, Farm Forestry Research Programme, PO Box 20412, Nairobi, Kenya. E-mail: nyamaikefriaf@form-net.com.

CAB International: A leading non-profit organisation specialising in sustainable solutions for agricultural and environmental problems. CABI Bioscience is its research wing, specialising in applied biological sciences for sustainable agriculture and environmental safety, and CABI Publishing produces materials on applied life sciences. Its headquarters are in the UK, but CAB International also has regional offices in Africa, SE Asia, Latin America, and the Caribbean. Contact details: Nosworthy Way, Wallingford, Oxon OX10 8DE, UK. E-mail: corporate@cabi.org or publishing@cabi.org; Web: www.cabi.org

Center for International Forestry Research (CIFOR): A global knowledge organisation based in Indonesia whose mission is to enhance the benefits of forests for people in the tropics. It operates through a series of decentralised partnerships with key individuals and institutions worldwide. One of its core objectives is to facilitate the transfer of knowledge across countries and to strengthen national capacities for research, to support the development of policies and technologies for the optimal use of forests and forest land. Contact details: PO Box 6596, JKPWB, Jakarta 10065, Indonesia. E-mail: cifor@cigiar.org; Web: www.cifor.cgiar.org

Consultative Group on International Agricultural Research (CGIAR): Founded in 1971, CGIAR is an association of public and private members, supporting a system of 16 Future Harvest Centers working in more than 100 countries, which are committed to the promotion of food security, the eradication of poverty, and the protection of the environment in the developing world. In particular, CGIAR advocates the use of innovative research (strategic and applied) and science-based approaches to address some of the world's most pressing developmental problems. Contact details: CGIAR Secretariat, The World Bank, MSN G6-601, 1818 H Street NW, Washington, DC 20433, USA. E-mail: cgiar@cgiar.org; Web: www.cgiar.org

European Tropical Forest Research Network (ETFRN): Created in 1991 to provide a focal point for information exchange and debate among organisations, institutions, scholars, and researchers in Europe with an interest in (sub)tropical forest research, the Network seeks to encourage the involvement of European research expertise in the conservation and wise use of forests and woodlands in tropical and subtropical countries. Services include an on-line database of European institutions involved in tropical, subtropical, and Mediterranean forest research, a question and answer service, an international calendar of relevant events, and *ETFRN News*, its quarterly newsletter. Contact details: c/o The Tropenbos Foundation, PO Box 232, 6700 AE Wageningen, The Netherlands. E-mail: ETFRN@iac.agro.nl; Web: www.etfrn.org

Food and Agriculture Organisation (FAO) – Forestry Department: The Forestry Department seeks to ensure the development of policies, strategies, and guidelines for FAO members, as well as providing relevant research analysis and advisory and technical services. It promotes national and international action for the effective conservation, sustainable management, and efficient use of forest and related resources as an integral element of land-use systems. Contact details. Viale delle Terme di Caracalla, 00100 Rome, Italy. E-mail: ftpp@fao.org; Web: www.fao.org/forestry/fo/STRUCT/en/struct_e.html

Forest Trees and People Programme (FTPP): FTPP has contacts worldwide who maintain a network to facilitate the sharing of information about community forestry activities. The central website also provides reviews and an ordering facility for many relevant publications from FAO and the Swedish University of Agricultural Sciences, which jointly run the programme with regional contacts. A quarterly newsletter is distributed free of charge in English, French, and Spanish. Contact details: FTP Network, SLU Kontakt, Swedish University of Agricultural Sciences (SLU), Box 7034, 750 07 Uppsala, Sweden. E-mail: ftp.network@kontakt.slu.se; Web: www-trees.slu.se

Intermediate Technology Development Group (ITDG): An international NGO which facilitates the use of technology to identify practical answers to poverty, especially in the area of agriculture. Based in the UK, ITDG also has offices in Bangladesh, Kenya, Nepal, Peru, Sri Lanka, Sudan, and Zimbabwe, through which it runs programmes designed to help poor communities to develop appropriate technologies in food production, agro-processing, energy, transport, small-enterprise development, shelter, small-scale mining, and disaster migration. To disseminate findings and lessons from its grassroots experiences, ITDG offers consultancy services, policy papers, and publishing and educational activities. Contact details: ITDG, Myson House, Railway Terrace, Rugby CV21 3HT, UK. Web: www.itdg.org

International Centre for Research in Agroforestry (ICRAF): Established in Nairobi in 1977, ICRAF is a non-profit research body seeking to alleviate poverty, improve food and nutritional security, and enhance environmental sustainability in the tropics. Supported by the CGIAR, ICRAF conducts strategic and applied research, in conjunction with national agricultural research systems. ICRAF also forms partnerships with a range of development institutions to facilitate the adoption of agroforestry practices by smallholder farmers, as well as policies and institutional innovations to promote sustainable and productive land use. Its work is based on five research and development themes, which include diversification and intensification of land use through domestication of agroforestry trees; soil-fertility replenishment in nutrient-depleted lands with agroforestry and other nutrient inputs; socio-economic and policy research to allow policies that will benefit small farmers; acceleration of impact on farms, and capacity and institutional strengthening through training and the dissemination of information. Contact details: PO Box 30677, Nairobi, Kenya. E-mail: ICRAF@cgiar.org; Web: www.icraf.cgiar.org

International Development Research Centre (IDRC): A public corporation created in 1970 to work with developing countries in identifying long-term solutions to the social, economic, and environmental problems that confront them. Relying on opportunities provided by science and technology, IDRC focuses on knowledge gained through research as a means of empowering the people of the South. To this end, the Centre funds the work of scientists working in universities, private enterprises, government, and non-profit organisations in developing countries, and supports regional research networks and institutions in the developing world. Contact details: 250 Albert Street, PO Box 8500, Ottawa, Ontario, Canada K1G 3H9. E-mail: info@idrc.ca; Web: www.idrc.ca/institution/general_index_e.html

International Institute for Environment and Development (IIED) – Forestry and Land Use (FLU) Programme: IIED's Forestry and Land Use Programme, established in 1984, seeks to improve forest-based livelihoods and land use on the basis of equity, efficiency, and sustainability, focusing on arenas of critical decision making. Within its four broad themes – policy management and institutional change; the promotion of sustainable forestry and land use; the tackling of inequality; and awareness of international initiatives – FLU collaborates with a broad range of partners worldwide. Its research work feeds into training courses and learning groups, support for advocacy coalitions and policy makers, informational materials, and tools with which to influence policy. Contact details: 3 Endsleigh Street, London WC1H oDD, UK. E-mail: mailbox@iied.org; Web: www.iied.org/forestry/index.html

International Institute of Rural Reconstruction (IIRR) Philippines: An NGO devoted to improving the quality of life of marginalised communities in Africa, Asia, and Latin America, IIRR relies on bottom-up, participatory, integrated strategies to overcome rural poverty. IIRR has its own field-based research activities in the Philippines (its central programme), as well as in its offices in Ecuador and Kenya, to test new technologies and rural development approaches. Contact details: Y.C. James Yen Center, Silang, Cavite, 4118, Philippines. E-mail: Information@iirr.org; Web: www.iirr.org (under construction)

International Union of Forest Research Organisations (IUFRO): An international network of forest scientists that seeks to encourage international co-operation in forestry and forest products research. Among other things, IUFRO promotes the use of science in the formulation of forest-related policies among individuals, organisations, and relevant decision-making bodies. Contact details: Seckendorff-Gudent-Weg 8, A-1131 Vienna , Austria. E-mail: iufro@forvie.ac.at; Web: iufro.boku.ac.at/iufro

Kenyan Forestry Research Institute (KEFRI): KEFRI was established to enhance the social and economic welfare of local people through user-oriented research and development in forestry and allied natural resources. To do this, KEFRI seeks to generate technologies for farm forestry, natural forests, drylands forestry, and forest plantations, as well as document and disseminate scientific information. Contact details: PO Box 20412, Nairobi, Kenya. E-mail: kefri@arcc.or.ke; Web: www.easternarc.org/pu/kefri_strategic_plan.html

Natural Resources Institute (NRI): A specialised institute at the University of Greenwich which provides training, research, consultancy, and advisory services to underpin sustainable development. Its main areas of work include livelihoods, environment, agricultural systems, and ecosystem management. NRI also plays a significant role in relation to institutional capacity building in developing countries, through subcontracting research and consultancy to its partners overseas. Contact details: Medway University Campus, Central Avenue, Chatham Maritime, Kent ME4 4TB, UK. E-mail: nri@greenwich.ac.uk; Web: www.nri.org

New Forests Project (NFP): A people-to-people, direct-action, grassroots programme created in 1982 in an effort to initiate reforestation and reduce deforestation in more than 120 developing countries. The project seeks to provide farmers, community organisations, and environmental groups with the training and information necessary to begin successful reforestation projects and to protect their watersheds. The main objective is to generate realistic alternatives to the harvesting of existing tropical forests, in order to protect the ecosystem from further erosion. Contact details: 731 8th Street, SE, Washington, DC 2003, USA. E-mail: icnfp@erols.com; Web: www.newforestsproject.com

Overseas Development Institute (ODI): A number of ODI's research programmes relate to agroforestry and on-farm research. The Forest Policy and Environment Group and the Rural Policy and Environment Group have produced a wealth of information on participatory research with farmers. Relevant papers include: *Institutional Development of Local Organisations in the Context of Farmer-led Extension: the Agroforestry Programme of the Mag'uugmad Foundation*, David Brown and Caroline Korte, 1997; *Organisational Roles in Farmer Participatory Research and Extension: Lessons from the Last Decade*, John Farrington, 1998; *Rethinking Approaches to Tree Management by Farmers*, Michael Arnold and Peter Dewees, 1998; and *From 'Tree-haters' to Tree Farmers: Promoting Farm Forestry in the Dominican Republic*, F. Geilfus, 1997. Contact details: 111 Westminster Bridge Road, London SE1 7JD, UK. E-mail: odi@odi.org.uk; Web: www.odi.org.uk

UK Agroforestry Forum: The UK Agroforestry Forum is an informal group of people with a common interest in agroforestry. While mostly academic and research-based, the forum is expanding to incorporate the views and insights of farmers, foresters, conservation agencies, and other practice-oriented groups. The Forum has set up a JISCMail mailing list to foster the development of agroforestry systems in temperate regions through discussion of research, technology transfers, and socio-economic and policy issues. The Forum holds an annual meeting to promote the dissemination of the latest research, developments, and practices in the area, which is complemented by its *UK Agroforestry Forum Newsletter*. Web: www.agroforestry.ac.uk

Wageningen University: Since it was founded in 1918, Wageningen University has become a leading educational and research centre in the plant, animal, environmental, agro-technical, food, and social sciences. Its objective is to develop and disseminate the scientific knowledge needed to supply sufficient, healthy food to meet world demand within an ecologically sound environment, and in a sustainable fashion.

Research is focused on four central themes: sustainable agricultural production chains; agrotechnology, nutrition, and health; nature development and conservation of natural resources, and spatial planning, environmental planning, and water management in rural areas. Contact details: Postbus 9101 6700 HB Wageningen, The Netherlands. E-mail: info@www.wag-ur.nl; Web: www.wur.nl

Winrock International: A non-profit organisation that works with people around the world to increase economic opportunity, especially in the rural sector, stimulate agricultural productivity, sustain natural resources, promote responsible resource management, and protect the environment. Winrock aims to match innovative approaches in agriculture, natural resources management, clean energy, and leadership development with the needs of its partners. By linking individuals and communities with new ideas and technology, Winrock seeks to increase long-term productivity, equity, and responsible resource management to benefit the poor. *Innovations*, its monthly newsletter, as well as other publications, are available on the Internet free of charge. Contact details: 30 Winrock Drive, Morrilton, AK 72110, USA. E-mail: mail@winrock.org; Web: www.winrock.org

Addresses of publishers

(addresses for organisations are listed under individual entries)

ACIAR
ACIAR House, Traeger Court, Fernhill Park, Bruce Act 2617, Australia.
E-mail: aciar@aciar.gov.au

Academic Press
525 B Street, Suite 1900, San Diego, CA 92101-4495, USA. E-mail: ap@acad.com

Agroforestry Research Trust
46 Hunters Moon, Dartington, Totnes, Devon TQ9 6JT, UK.
E-mail: mail@agroforestry.co.uk

Anmol Publications
Ansari Road, Darya Ganj, New Delhi, India.

Ashgate Publishing Ltd.
Gower House, Croft Street, Aldershot, Hampshire GU11 3HR, UK.
E-mail: info@ashgate.com

CABI
Wallingford, Oxon OX10 8DE, UK. E-mail: orders@cabi.org

CRC Press
2000 NW Corporate Blvd, Boca Raton, FL 33431, USA. E-mail: orders@crcpress.com

Earthscan Publications Ltd.
120 Pentonville Road, London N1 9JN, UK. E-mail: earthinfo@earthscan.co.uk

FAO
Viale delle Terme di Caracalla, 00100 Rome, Italy. E-mail: FAO-HQ@fao.org

ICRAF
PO Box 30677, Nairobi, Kenya.

International Service for National Agricultural Research
PO Box 93375, 2509 AJ The Hague, The Netherlands. E-mail: isnar@cgiar.org

IT Publications
103-105 Southampton Row, London WC1B 4HL, UK. E-mail: itpubs@itpubs.org.uk

Kluwer Academic Publishers
PO Box 17, 3300 AA Dordrecht, The Netherlands. E-mail: kluweronline@wkap.nl

Lynne Rienner
1800 30th Street, Suite 314, Boulder, CO 80301, USA. E-mail: questions@rienner.com

Magraf Verlag
PO Box 1205, 97985 Weikersheim, Germany.

National Agroforestry Center
North 38th St. & East Campus Loop, UNL-East Campus, Lincoln, Nebraska 68583-0822, USA.

Oxford & IBH Publishing Co.
Pvt. Ltd. 66 Janpath, New Delhi 110 001, India . E-mail: Oxford@nda.vsnl.net.in

Permanent Agriculture Resources
PO Box 428, Holualoa, Hawaii 96725, USA. E-mail: training@agroforester.com

Sage Publications, India
M-32 Market, Greater Kailash-1, New Delhi-110024, India.
E-mail: sageind@giasdl01.vsnl.net.in

University of Texas Press
PO Box 7819, Austin, Texas 78713-7819, USA. E-mail: utpress@uts.cc.utexas.edu

University of California Press
2120 Berkeley Way, Berkeley, CA 94720, USA. E-mail: askucp@ucpress.edu

Yale University Press
PO Box 209040, New Haven, CT 06520-9040, USA.
E-mail: custservice.press@yale.edu

Zed Books
7 Cynthia Street, London N1 9JF, UK. E-mail: general@zedbooks.demon.co.uk

Index

Acacia spp., planted in woodlots 41
aerial surveys, Kenya, show increase in
    woody biomass 59, 164
Africa
    sub-Saharan
        extension services 57
        low quality and quantity of
        livestock feed resources 107
    see also Kenya; Malawi; Uganda;
        Zambia
Africa, Southern
    development trends and
        agroforestry opportunities 36–8
    facilitating the wider use of
        agroforestry for development
        35–55
        approaches to accelerating
        impact: agroforestry as a
        learning tool 4-5
        developing agroforestry options
        38–43
        some lessons and preliminary
        conclusions 50–4
    key farming constraints 35–6
    offering alternative practices and
        tree species important 157
    research assessing market demand
        for indigenous fruit 162
    soils and crops 35
    women as participating farmers 159
African Highlands Initiative approach,
    Kenya 147–8
    farmer research committees 148
    resource persons' subcommittee of
        village committee 148

African Network of Agroforestry
    Education (ANAFE), goal of 9
agricultural innovations, how far are
    they acceptable 2
agriculture, and public-sector
    extension 56
agroforestry
    analysis of adoption 3–4
    assessing dissemination potential
        of partner organisations 166
    classed as preventive innovations
        43
    developing technology options,
        southern Africa 38–43
        annual relay cropping of trees
        40–1
        fodder banks 41
        improved fallows 39
        mixing coppicing trees and
        crops 39–40
        planting indigenous fruit trees
        42–3
        rotational woodlots 41–2
    diagnosis of potential, Yucatán
        18–19
        interviews and agenda
        formulation 18
    effective dissemination through
        community organisations 76–7
    impact over different temporal
        scales 5
    innovations
        factors affecting adoption of 4
        linking to development,
        Southern Africa 43–4

large-scale, has to be affordable
80–1
as a learning tool 44–50, 53–4
monitoring and evaluation: key
element in the learning process
49–50
pilot development projects
47–9
scaling out through partners
44–7
and local policy makers 163
mainly provides substitutes for
purchased inputs 162
meeting multiple needs 48
multifaceted nature of scaling up
167–9
must be suitable and cheap 118
problems inhibiting adoption of,
Uganda 70
realising the potential of 1–14
fundamentals of adoption and
impact 2–5
institutional change: towards a
research and development
continuum 6–8
strategy for scaling up: crucial
areas of investment and
intervention 8–11
agroforestry impact, timeframe for
51–2
agroforestry innovations, essential
elements for scaling-up 8, 9
basic education institutions 10
community organisations 10
extension and development
organisations 11
higher education institutions 9
policy makers 8
product marketing systems 10–11
research institutions 11
seed-supply systems 10
agroforestry products, better markets
generate income for poor
households 10–11
agroforestry research
importance of 11–12
scaling up the benefits of 156–7

building local capacity 159–61
farm-centred research and
extension 158–9
germplasm 161
knowledge and information
sharing 164–5
learning from successes and
failures 163–4
market options 161–2
policy options 162–3
research challenges on scaling
up 167–9
strategic partnerships and
facilitation 165–7
technical options 156–7
agroforestry systems, participatory
design of 15–23
impact of participatory research
and empowerment of farmers
22–3
stages of design 16–22
adaptation of participatory
methods 20–2
design and implementation of
trials 19
diagnosis of potential of
agroforestry 18–29
establishing farmer groups
16–18
evaluation of trials 19–20
agroforestry technologies
dissemination of 43–4
scaling up, adoption and impact of,
western Kenya 136–55
agroforestry trials
adaptation of participatory
methods 20–2
improved fallows, gaining
experience with new species 21
design and implementation 19
evaluation 19–20
of cover legumes in fruit-timber
agroforestry 20, 21
alley farming, Cameroon 4
Alnus acuminata 76
Alternatives to Slash-and-Burn
Consortium (Cameroon) 5

Australia, Landcare movement 132–3
Azadirachta indica (neem tree) 41

backyard gardening 129
Bactris gasipaes (peach palm) 25
    under-utilised food crop 26
biomass transfer 1, 12n
    of *Tithonia diversifolia* 137
boundary planting 76

Calliandra calothyrsus, fodder shrub
    107
    better seed germination after
        longer soaking 112–13, 165
    planted as a dense hedge 137
    poor performer in central Kenya
        157
    research on 108–9
        an effective substitute for dairy
            meal 109
        grown in hedges, on contour
            bunds or intercropped 108
        seedlings raised in nurseries
            109, 110–11, 112
Calycophyllum spruceanum 25, 157
    differs according to bark colour 26,
        27
    diverse provenances, to be
        managed for conservation/seed
        production 30
    on-farm provenance trials 27–9
    in progeny trials 30–1
    provenance and progeny trials can
        be converted to seed orchards
        32
CARE
    Agroforestry Project 139
        group resource persons (GRPs)
            139
        worked with women's groups
            and schools 139
    TRACE programme 149–50
        benefits associated with 150
CBOs *see* community-based
    organisations (CBOs)
churches/social groups, good entry
    points for extension 151

Claveria, Mindanao
    change in tillage system 119–20
    changing attitudes of farmers,
        policy makers, local
        government and land owners
        129–30
    Claveria Landcare Association 123
        chapters and sub-chapters 126
    contour-hedgerow concept popular
        119
    contour hedges (SALT) 118–119
    effects of heavy rainfall on sloping
        fields 118
    farming based on two crops of
        maize per year 118
    Landcare groups
        based in sub-villages 123
        successfully extended
            conservation farming 124
    trash bunds and natural vegetative
        strips (NVS) 119–20
communities
    action needed to address watershed
        degradation 82
    feedback through village elders and
        social group leaders 63
    prerequisite for successful action
        74
    should represent broad range of
        conditions 22
    understanding the complexity of 48
    wish for pluralistic extension
        approach 67
    see also local communities
community action
    Katagata watershed, Kabale District
        75
        agroforestry innovations
            available to Kyantobi farmers
            75–6
community development,
    achievement of 153–4
community organisations
    effective, disseminate agroforestry
        information/systems 76–7
    provide Mexican farmers with
        forum for discussions 22

community-based organisations
(CBOs) 110, 151
follow-up and support crucial for 154
used in agroforestry projects 138
conservation technologies
adopted by Landcare members
128–9, 160–1
appropriate sites on sloping land,
Claveria 124
minimum-tillage or ridge-tillage
systems 129
Consortium for Increasing Farm
Productivity in Western Kenya
152–3
backed up by KARI/KEFRI/ICRAF
pilot projects 153
Consultative Group for International
Agricultural Research (CGIAR) 1
contour bunds
for land-shaping, Nagaland 88
low adoption rate 163–4
for *Tithonia* hedging 142
contour hedgerows 2, 137
important part of soil conservation,
South-East Asia 117
for soil conservation, Kabale
District 71, 75–6, 77, 160
unpopular in the Philippines
117–20
Crotolaria spp., sown under
established crops 40
Cupressus lusitanica 61

dairy cows, tree legumes provide
protein supplement for 1–2, 109–12
decentralisation
of government functions, Uganda
71, 72–4, 82
decision-making, sharing knowledge
and information critical to 164–5
deforestation 24, 35
Nagaland, caused by swidden
agriculture 84
Desmodium intortum 112
development 6, 7
linking agroforestry innovations to
43–4

development indicators, Southern
Africa 37
dissemination (of agroforestry)
as a community responsibility
137–8
and on farm testing 84–106
pilot projects 1, 47–9, 61–4
through farmer groups 114
through community organisations
76–7
using village committees 144
drought, and seedling mortality
111–12, 113

'early adopters' 52
environmental degradation, Southern
Africa 35–6
erosion control, Nagaland
runoff blockades 93
traditional
preferred 103
used before NEPED 102
Eucalyptus spp. 157
extension, participatory 158–9
extension contacts, important during
early farmer experimentation 11
extension services 166
Kenya 58–9, 146
access to reinforced 60
blanket recommendations not
encouraged 61
conventional channels,
effectiveness assessed 59
crucial in letting pilot projects
reach full potential 65
programmes with strong
emphasis on bottom-up
approach 68
raising of morale 60
western, retraining to facilitate
farmer participation 142
often weak 137
sub-Saharan Africa 57
extension staff, costs of collaboration
with 51, 166

fallows
  improved 21, 39, 157
    provide fuelwood and stakes 137
  leguminous 1, 12n
  short-duration 137
  short-term 2
  see also forest fallow; tree fallows
farm forestry, success depends on
  number of free seedlings 91
farmer groups
  differences between indigenous
    and immigrant farmers 17–28
  Embu district, dissemination
    through 114
  establishment of, Yucatan
    Peninsula 16–18
  types of groups, strengths of 17
  evaluate component species for
    agroforestry systems 19, 20
farmer preference surveys 158
farmer research committees, Kenya
  148
farmer-to-farmer group training 46,
  165
farmers
  as agents of change 56
  alternative practices and tree
    species for problem solution
    important 157
  Claveria
    role of in Landcare system 125
    shown successful technologies
      and organisational methods
      124–5
  Embu District, Kenya
    aim to extend number using
      Calliandra as dairy feed 109–12
    big demand for fodder shrubs
      113
    training in nursery
      establishment and seed
      production 110–11, 112
    enabled to analyse and plan range
      of options and solutions 82
    evaluation criteria 19
    faster delivery of high-quality
      planting material to 31–2

field exchange visits described as
  inspirational 145
implementing agroforestry trials
  19
increased role in non-farm
  participatory research 4–5
interactions with research and
  extension services 57
Kenya, farmer field school 146
as key change agents 52–3
learning from mapping exercise 77,
  78, 79
low-income, discount long-term
  benefit of trees 5
Nagaland
  jhum (swidden) agriculture 85
  realism of 104
and principles of tree
  domestication 25–6
promotion and facilitation of
  innovation adoption among 5
selecting improved tree-planting
  material with 26–30
consideration of genetic
  diversity essential 30
on-farm provenance trials
  27–9, 32
opportunities for selection and
  improvement 26–7
'systematic collection' strategy
  27
useful information about
  selection criteria 26, 27
selection of for test plots, Nagaland
  90
should have vested interest in
  conserving tree genetic
  resources 33
small, meeting challenge of market
  liberalisation and deregulation
  56
value of in extension work 62
western Kenya, problems of 136
farmers' organisations, Landcare,
  Claveria 126
farmers' trials, used as demonstration
  plots 22

Finnish International Development
Agency (FINNIDA) 56–7
fodder banks 41
aim to increase income of
smallholder dairy farmers 41
Calliandra, economic analysis of 112
Kenya, tree legumes in 1–2, 112
fodder shrubs, scaling up the use of in
central Kenya 107–16, 164
achievements and impact 109–12
aims apart from transfer of
knowledge 109–10
consortium of partners needed for
promotion 115
diversification in shrub species
advisable 112, 114
factors contributing to success
113–14
monitoring, farmer innovation and
feedback 112–13
problems encountered 113
remaining challenges 114–15
research on fodder shrubs 108–9
food insecurity, Southern Africa 35
food security
and agroforestry 37
attainment of 136
through improved fallows 39
forest fallow
Nagaland 84
restores fertility, Peruvian Amazon
24
forest fragmentation 24
fruit trees
for home consumption 76, 128
indigenous, planting of, Southern
Africa 42–3
limitations 43
objectives of planting 42–3
priority species 42
require a different approach 81

genetic diversity, important
consideration in selecting tree
populations for cultivation 30
genetic resources, need to use
sustainably 33

germplasm 60, 161
support for decentralised grassroots-
level production 47
Gliricidia sepium, for coppicing 39
Gmelina arborea 129
grassroots participation, importance of
recognised 88–9
Grevillia robusta 61, 76, 137, 157
Guazuma crinita 25
on-farm provenance trials 27–9
in progeny trials 30–1
provenance and progeny trials,
conversion to seed orchards 32

ICRAF *see* International Centre for
Research in Agroforestry (ICRAF)
ILRI *see* International Livestock
Research Institute (ILRI)
impact assessment 5
India *see* Nagaland
indigenous knowledge, and tree
domestication, Peru 25–6, 164–5
indigenous people, often vulnerable to
outside world 87–8
Inga edulis 25
valued more by women 26
innovation-decision period, length of 52
innovation-decision process, stages of
2–3
innovations, types of impact from
adoption of 5
insect pests, and tree fallows 39
Integration of Tree Crops into
Farming Systems, Kenya 47
International Centre for Research in
Agroforestry (ICRAF) 107
better methods of forecasting
germplasm needs 10
Claveria, Mindanao
help given to an adjacent
municipality 132
and the Landcare approach
123–34
research on contour-hedgerow
technology 117
support for dissemination
activities 123

creation of development division 7
development of training-of-trainers
strategy for Landcare facilitators
132
direct engagement in the
development process 12, 36
evaluating/disseminating
agroforestry technologies with
Kenyan partners 136–7
farmer-participatory research 1
Farmers of the Future programme
10
institutionalising and
strengthening partnerships
152–3, 165, 166
investment in process-oriented
research 1
lessons learned and challenges
ahead using development
partners 152
on-the-ground partnerships with
development organisations 7–8
getting the right partnerships
50–1
government partners are
special cases 51
pillars of research and
development 6
pilot dissemination projects 1,
47–9, 61–4
placing market research in
mainstream research and
development 11
proactive engagement in
development process, benefits
of 6–7
regional strategic planning
exercise, Southern Africa 37–8,
38
research and development activities
integrated in the field 7
scaling out with partners, Southern
Africa 44–7
collaborating partners 45–6
networkshops, main
instrument for collaboration
46, 167

raising awareness of
stakeholders 36
role as facilitator 44–5
strengthening grassroots
capacity 46
way forward for development
division 8
working to counter deforestation-
forest fragmentation-soil
degradation-poverty cycle 25
Zambia project, building capacity
in other organisations 167
International Institute of Rural
Reconstruction (IIRR), defined
scaling up 156
International Livestock Research
Institute (ILRI) 107, 108

Kabale, Uganda see Katagata
watershed, Kabale District
KARI see Kenya Agricultural Research
Institute (KARI)
KARI/KEFRI/CARE pilot project
approach
introduced high-value tree species 143
lessons learned and challenges
ahead 144–5
making dissemination a
community responsibility
137–8
pilot villages acting as training
points 143
scaling up through activities of
other development partners
145–51
adaptive research, KARI-Kissi
approach 146
African Highlands Initiative
approach 147–8
community-based
organisations 151
educational institutions 151
government ministries and
projects 146
NGOs 149–50
Participatory Learning Action
Research village project 146–7

Tropical Soil Biology and
Fertility interactive learning
project 148–9
village committee approach 140–3
achievements and impacts
142–3
community-based
dissemination methodology
140, 141
mobilisation of collective action
142
Katagata watershed, Kabale District
climate, topography, and soils 71
contour hedgerows seen as most
suitable innovations 75–6
farmers beginning management of
74
involvement in agroforestry 77
Kyantobi farmers
identified other agroforestry
innovations to try 76
study tour to other research
sites 75–6
problem of runoff after heavy rains
74–5
effective control requires
community action 75
smallholder agriculture 72
Kenya
central, scaling up the use of fodder
shrubs 107–16
Calliandra important for dairy
goats 112
potential collaborating
organisations identified 110
study area described 108
Embu District 107
farmers experienced in feeding
Calliandra to dairy cows 110
farmers set up *Calliandra*
nurseries 109, 110–11, 112
tree legumes in fodder banks
1–2, 112
extension-service focus on farmers
reaching other farmers 138
fodder shrubs slow to reach many
dairy farmers 114

Nakuru and Nyandarua intensified
forestry extension project 57–9,
159
enhanced implementation of
conventional service delivery
59–61
piloting participatory extension
approaches 61–4
National Agriculture and Livestock
Extension Project (NALEP)
67–8
will incorporate more
participation in decision
making by stakeholders 68
National Soil and Water
Conservation Programmes,
impact analysis of 67
scaling up participatory
agroforestry extension 56–69
current trends in extension and
expected components of
extension approaches 56–7
linking pilots to policy 65–8
programme run by
Government and FINNIDA
56–7
Soil and Water Conservation
Programme 146
tree-felling permits made
redundant 60, 162–3
western
community participation and
on-farm testing 142, 160
Consortium for Increasing
Farm Productivity in Western
Kenya 152–3, 167
improved food availability 143
scaling up adoption and impact
of agroforestry technologies
136–55
Kenya Agricultural Research Institute
(KARI) 136
Agricultural Technology and
Information Response
Initiative (ATIRI) 145
KARI/KEFRI/CARE pilot project
approach 137–8, 140–3

testing fodder shrubs 107
Kenya Forestry Research Institute
    (KEFRI) 107, 136
Kenya Woodfuel and Agroforestry
    Project (KWAP) 138–9
    umbrella development groups
        138–9
    used A–B–C framework 138
KWAP *see* Kenya Woodfuel and
    Agroforestry Project (KWAP)

land degradation, sustainable
    agroforestry solutions for *see*
    biomass transfer; contour
    hedgerows; fallows; fodder banks;
    natural vegetative strips (NVS)
land tenure, Nagaland 86, 104
land-shaping 90
    concept of 88
Landcare movement, the Philippines
    10, 160–1
    conservation farming technologies
        adopted 128–9, 160
    different models for scaling up 131
    framework for development of the
        approach 130, 131
    impacts and scaling up 129–34
    organisational structure of 128
    provision by Landcare approach 130
    steps involved 124–7, 128
    support from external donor
        agencies 127
landholding, Yucatán Peninsula 16
leguminous fallows 1, 12n
Leucaena spp. 21, 41
Leucaena trichandra 109, 112
live hedges 2
LLP *see* local-level planning (LLP)
local agricultural research councils
    (CIALs) 22
local communities
    benefit from technical backup 77
    empowerment
        as change agents 168
        importance of 10
    pilot project, Nyandarua District,
        key observations 62–3

local government
    Claveria
        support to Landcare Association
            125–7
    Uganda 73, 77–80
        sub-counties, suitable for
            community action 77–8, 80
local-level initiatives (LLIs)
    lessons learned so far 67
    SIDA-Kenya development
        programme 66
local-level planning (LLP) 66
    experience with 61–2
    pilot project in Nyandarua District
        61–2
        community action plan
            developed 61
        results shared with policy makers
            65–6

maize crops, green manure within or
    between 21–2
Malawi, Zomba, hypothesis,
    conservation farming on steep
    slopes (pilot scheme) 48–9, 164
market options, weakest element in
    scaling up 161–2
marketing 10–11, 168
    of tree crops, Southern Africa 47
Mexico, farmer participatory methods
    15–23
    being used in Government
        development projects 22–3
    impact of participatory research
        and the empowerment of
        farmers 22–3
    stages of participatory agroforestry
        system design 16–22
    Yucatán Peninsula, opportunities
        for agroforestry in 16
Mindanao, Philippines *see* Claveria,
    Mindanao
monitoring
    importance of 81
    informal 112–13
monitoring and evaluation
    by three types of actor 48–9, 164

enhances learning among
stakeholders 163–4
key element in the learning process
49–50
in the Landcare programmes 127
practical approach founded on
three pillars 50
and scaling up 105
monoculture, maize, depletes soil
fertility 36
Morus alba (mulberry) 109
Mucuna 21

NAFRP *see* National Agroforestry
Research Project (NAFRP), Kenya
Naga people, divided into many tribes 86
Nagaland Environmental Protection
and Economic Development
(NEPED) Project
Canadian funding for 89
design of 90–3
choice of denser planting,
benefits of 92, 165
contour bunds not liked,
modified land-shaping allowed
91
food crops to be integrated with
timber trees 91
gender component added after
1996 91
local practices, ecological
insights and innovations seen
and disseminated 91–3
'search and find' philosophy 90
test plots for experimentation
and dissemination 90, 92
farmer practices 100–2
heavy reliance on teak and
Gmelina arborea 101
land shaping in *jhum* fields not
extensive 102
replicate plots grow more of
fewer species 101
selective weeding and re-growth
of valuable species 101
impact on community of scaling up
96–100

concern that enthusiastic tree
planting might increase
deforestation 98–9, 104
village elders asked to assemble
information about tree planting
96–8
lessons 102–5
successful in stimulating
replication 103
most commonly planted species
93, 94
original impetus for arose locally
100
replication plots often differ from
basic concept 103–4
scaling up the project 93–5
early evidence of extensive tree
planting 95
scheme diversity fundamental to
successful scaling up 104
survey to measure extent of scaling
up 95–6
Nagaland, India
on-farm testing and dissemination
of agroforestry 84–106
efforts to counteract expanding
jhum and deforestation 88
governance foundation of
society in villages 87
land set aside for forest reserves
in each village 98
land usually locally controlled
86
land-shaping and tree-planting
experiments 88–9
land-use decisions taken jointly
within the village 94
land-use systems 85–7
village elders do not see tree
planting affecting food supply
99
village resolutions to plant trees
100
Nakuru and Nyandarua Intensified
Forestry Extension Project (Kenya)
57–9
components of 57–8

enhanced implementation of conventional service delivery 59–61
extension approaches and agroforestry technologies matched to specific requirements 61
implemented by Forestry Extension Services Division 57
lessons learned and recommendations 63–4
piloting participatory extension approaches 61–4
experience with local-level planning 61–2
key observations 62–3
lessons learned and recommendations 63–4
logistical difficulties in implementation 63–4
schools approach 58
Training and Visit system 58
NALEP *see* National Agriculture and Livestock Extension Project (NALEP)
Napier grass (*Pennisetum purpureum*) 108
National Agriculture and Livestock Extension Project (NALEP) 67–8, 146
National Agroforestry Research Project (NAFRP), Kenya 107
National Institute for Natural Resources (INRENA) 32
natural resource management, more effective, Uganda 70–83
agroforestry innovations available 75–6
community organisations 76–7
demand-driven approach 74–5
Katagata watershed 71–2
local government 77–80
minimum-input strategies 80–1
policy framework 72–4
natural vegetative strips (NVS) 2, 119–20, 128
NEPED *see* Nagaland Environmental Protection and Economic Development (NEPED) Project

networkshops, Southern Africa, 46, 167
NGOs
western Kenya, working with ICRAF 149–50
CARE programme 149–50
nurseries
Calliandra nurseries 109, 110–11, 112, 113
fruit and timber trees, Claveria 124, 129
NVS *see* natural vegetative strips (NVS)

on-farm (participatory) research *see* participatory (on-farm) research
on-farm surveys, of woody biomass 59
overlogging 24

participation
building grassroots capacity in situation analysis 50
using focus-group discussion 59
participatory design, of agroforestry systems 15–23
participatory forest management, Nagaland, practical steps to system improvement 89
Participatory Learning Action Plan (PLAR), Kenya 146–7
development of village action plans 147
participatory methods
adaptation to different circumstances 20–2
improved fallows 21
should be used in monitoring and evaluation 50
participatory (on-farm) research 4–5, 11
impact of and empowerment of farmers 22–3
should cover range of ecological, social, and economic conditions 22
southern Africa 43
participatory rural appraisal (PRA), use by local-level planning 61
partnerships, and scaling out/up 44–7, 165

Peruvian Amazon 161
  accelerating delivery of high-quality
    planting material 31–2
  adoption of tree-domestication
    methodology 32
  Aguaytía watershed
    chosen for on-farm provenance
      trials 27
    G. crinita much taller than other
      provenances 28
    targeted collections for progeny
      trials 31
    variation in wood density, *C.*
      spruceanum 28, 157
  demonstrating risk of poor tree
    adaptation to farmers 30–1
  dependency on trees 24
  diversity and quality of trees
    declining 24
  ideal provenance for timber and
    energy 28–9
  lessons learnt 33
  principles of farmer-driven tree
    domestication 25–6
  species chosen for
    domestication projects 25
  selecting improved tree planting
    material with farmers 26–30
Philippines, southern
  Contour-hedgerow systems 2
  Human Ecological Security (HES)
    programme 127
  Landcare movement 10, 123–34,
    163
    evolution of and innovative
      extension strategy 122–3
    evolving components of a
      successful conservation
      farming system 120–2
    National Strategy for Improved
      Watershed Resources
      Management, incorporation of
      Landcare approach 132
Philippines, the, Landcare experience
  117–35
phosphorus fertilisers 137
  use of reactive phosphate rock 143

policy makers 8, 65–6, 163, 169
poor people 159, 169
population growth
  Nagaland, effect on *jhum* cycles 85–6
  Southern Africa 35, 36–7
principal component analysis 28
progeny trials, Peruvian Amazon 30–1,
  32

relay cropping, of trees 40–1
research 162
  on *Calliandra calothyrsus* 108–9
  farmer-centred 158
  farmer-participatory 1, 4–5
    see also participatory (on-farm)
    research
  publicly funded, addressing food
    insecurity, poverty, and
    environmental degradation 6
research and development 5
  could be carried out and managed
    by Landcare groups 133
  integrated at ICRAF 6–8
  should involve farmers at all stages 81
research institutions, demand-driven
  and impact-oriented, needed 11
rotational woodlots 41–2, 76
  limitations 42
  main objectives 42
  potential production from 42

Sahel region, West Africa, use of live
  hedges to protect dry-season
  market gardens 2
SALT *see* sloping agricultural land
  technology (SALT)
Sapium ellipticum 109
savannah woodland eco-zone
  (*miombo*) 35
scaling out
  allows ICRAF only limited direct
    assessment of impact 51–2
  through partners 44–7
scaling up 82
  adoption and impact of
    agroforestry technologies,
    western Kenya 136–55

of agroforestry, must be cheap 80
defined 156
enabling policy environment
    critical 162–3
field practitioners minimise
    tension or conflict 158
of Landcare approach 129–34
of participatory agroforestry
    extension in Kenya 56–69
research challenges on 167–9
scaling up benefits of agroforestry
    research 156–70
    research challenges on scaling
    up 167–9
in the use of fodder shrubs in
    central Kenya 107–16
schools 139
useful in reaching the community
    151
venues for community-focused
    training 58, 59
seed
    Calliandra
        better germination after longer
        soaking 112–13
        commercial production and
        distribution 114
    few controls on sources for tree
        planting 30
    producing, distributing, marketing
        168
    testing questions of adaptation
        30–1
seed nurseries
    should be near water 113
    see also tree nurseries
seed orchards 32
seed production, training in 110–11,
    112
Senna siamea 41
Sesbania macrantha 40
Sesbania sesban
    for fallowing 39
    for relay cropping 40
SIDA *see* Swedish International
    Development Co-operation Agency
    (SIDA)

slash-and-burn agriculture, Peruvian
    Amazon 24
sloping agricultural land technology
    (SALT), Philippines 118–19
SLP *see* Systemwide Livestock
    Programme (SLP)
social change, through community soil
    and watershed conservation 49
solid-waste management 129
Southern Africa *see* Africa, Southern
Southern Africa Development
    Community (SADC) 36
sustainability, of Landcare movement,
    significant concerns about 133–4
sustainable management, of natural
    resource base in Nagaland, aim 90
Swedish International Development
    Cooperation Agency (SIDA),
    building on local-level initiatives,
    Kenya 66–7
swidden agriculture, Nagaland 84, 85,
    98–9
    cycles shortened 99
    integration of trees into 100, 102–3
    variations within systems 86
Systemwide Livestock Programme
    (SLP) 107, 109

technical facilitators, Claveria, role of
    in Landcare system 125
technical options *see* technology
    options
technology, needs development/trials
    on representative farms 64
technology options
    agroforestry
        aims to replenish soil fertility
        and aid food security 38–43
        take-off stage 52
    scaling up use of 156–7
technology transfer 156
    not always satisfactory 47
Tephrosia vogelii
    for annual relay cropping 40
    for tree fallows 39
Tithonia diversifolia, biomass transfer
    of 137

top-down solution, ineffective or damaging 87
Training Resource Persons in Agriculture for Community Extension (TRACE: CARE) 149–50
  benefits associated with 150
tree biodiversity, Nagaland, agroforestry a means of modifying traditional practices 84
tree domestication, farmer-driven, principles of 25–6
  document farmers' knowledge of variation within a species 25–6
  identify farmers' preferences 25
  men and women may value different species 26
tree domestication, practices, conservation function 30
tree domestication methodology, adoption of 32
tree fallows 39, 90
  factors affecting adoptability of 4
tree improvement
  alternative approach involving farmers 32
  traditional approach, too time-consuming 31
tree nurseries
  for fruit and timber trees, Claveria 124
  see also seed nurseries
tree planting, CARE project worked well 139
tree seed
  high-quality, disappearing, Peruvian Amazon 24–5
  limited availability of 40
tree-felling permits, Kenya, declared redundant 60, 162–3
trees
  agroforestry, participatory domestication of 24–34
  annual relay cropping of 40–1
  coppicing of and crops 39–40
  diversity of important 157
  and fruit trees, planting by Landcare members 128

high value, planted in western Kenya 143
leguminous, fast-growing 107
Nagaland
  expected benefits of planting 102
  planting for enriched fallow 90
  timber trees seen as additional crop 99
  nitrogen-fixing, fallowing with 39
Tropical Soil Biology and Fertility interactive learning project (Kenya) 148–9
  aims of 148–9

Uganda
  effects of decline in forest and plantation reserve 70
  Local Governments Act of Uganda, decentralisation programme 72
  managing watershed resources, successful sustainable community-based approaches 70
  more effective natural resource management 70–83
umbrella development groups 138–9

village committees, KARI/KEFRI/CARE pilot project approach 160
  aim 140
  awareness creation 144
  effective in disseminating technology 144
  forming the committee 140–1
  individual groups generally more active 144
  informal organisations, usefulness of 141
  purpose of 140
  roles and responsibilities not always clear 144–5
  study tours not always profitable to other farmers 145
  trained as necessary to improve performance 142

village development boards, Nagaland
90
  many plan to use funds for tree
    planting 99

women 159, 169
  addressing special needs of in
    agroforestry 53
  in agroforestry schemes 49
  included in farmer groups, Embu
    District–Kenya 111
  as members of farmer groups 16
  in NEPED project 91
    only limited success 105, 159
  using agroforestry technologies
    142
wood density, variations in
  C.spruceanum 28, 157
woodlots *see* rotational woodlots

Zambia
  networkshops 167
  short-rotation improved fallows
    germplasm for 161
    restoring soil fertility 2
  testing of improved fallows 43

Development in Practice Readers

Development in Practice Readers draw on the contents of the acclaimed international journal *Development in Practice*.

'The great strength of the Development in Practice Readers is their concentrated focus. For the reader interested in a specific topic ... each title provides a systematic collation of a range of the most interesting things practitioners have had to say on that topic. It ... lets busy readers get on with their lives, better informed and better able to deal with relevant tasks.'

(Paddy Reilly, Director, Development Studies Centre, Dublin)

The series presents cutting-edge contributions from practitioners, policy makers, scholars, and activists on important topics in development. Recent titles have covered themes as diverse as advocacy, NGOs and civil society, management, cities, gender, and armed conflict.

There are two types of book in the series: thematic collections of papers from past issues of the journal on a topic of current interest, and reprints of single issues of the journal, guest-edited by specialists in their field, on a chosen theme or topic.

Each book is introduced by an overview of the subject, written by an internationally recognised practitioner, researcher, or thinker, and each contains a specially commissioned annotated list of current and classic books and journals, plus information about organisations, websites, and other electronic information sources – in all, an essential reading list on the chosen topic. New titles also contain a detailed index. *Development in Practice Readers* are ideal as introductions to current thinking on key topics in development for students, researchers, and practitioners.

For an up-to-date list of titles available in the series, contact any of the following:

- the Oxfam Publishing website at www.oxfam.org.uk/publications
- the *Development in Practice* website at www.developmentinpractice.org
- Oxfam Publishing by email at publish@oxfam.org.uk
- Oxfam Publishing at 274 Banbury Road, Oxford OX2 7DZ, UK.

'This book [Development, NGOs, and Civil Society] will be useful for practitioners seeking to make sense of a complex subject, as well as for teachers and students looking for a good, topical introduction to the subject. There is a comprehensive annotated bibliography included for further exploration of many of the issues.'

(David Lewis, Centre for Civil Society at The London School of Economics, writing in *Community Development Journal* 36/2)

Development in Practice

'A wonderful journal – a real "one stop must-read" on social development issues.'

(Patrick Mulvany, Intermediate Technology Development Group, UK)

Development in Practice is an international peer-reviewed journal. It offers practice-based analysis and research on the social dimensions of development and humanitarianism, and provides a forum for debate and the exchange of ideas among practitioners, policy makers, academics, and activists worldwide.

Development in Practice challenges current assumptions, stimulates new thinking, and seeks to shape future ways of working.

It offers a wide range of content: full-length and short articles, practical notes, conference reports, a round-up of current research, and an extensive reviews section.

Development in Practice publishes a minimum of five issues in each annual volume: at least one of the issues is a 'double', focused on a key topic and guest-edited by an acknowledged expert in the field. There is a special reduced subscription for readers in middle- and low-income countries, and all subscriptions include on-line access.

For more information, to request a free sample copy, or to subscribe, write to Oxfam Publishing, 274 Banbury Road, Oxford OX2 7DZ, UK, or visit: www.developmentinpractice.org, where you will find abstracts (written in English, French, Portuguese, and Spanish) of everything published in the journal, and selected materials from recent issues.

Development in Practice is published for Oxfam GB by Carfax, Taylor and Francis.

'Development in Practice is the premier journal for practitioners and scholars in the humanitarian field who are interested in both practical insights and academic rigour.'

(Joseph G Block, American Refugee Committee, USA)